Advance Trends in Aquaculture Practices

THE EDITORS

Dr. A. Balasubramanian obtained his M.F.Sc., degree from Tamil Nadu Veterinary and Animal Sciences University and Ph.D from Central Institute of Fisheries Education, ICAR, Mumbai, specialized in Fishing Gears and Fisheries Resource Management. He is Professor and Head, Department of Fisheries Resource Management, College of Fishery Science, Sri Venkateswara Veterinary University, Muthukur, Nellore District, Andhra Pradesh, India. In addition, he served as Head of Research Station and College and University Head for the discipline of Fisheries Economics and Statistics and Academic Council member in the University. He teaches and guides the BFSc., MFSc., and Ph.D scholars in the discipline of Fisheries Resource Management at this Institute.

He has published more than 30 research papers in National and International journals and 5 training manuals and one book as author and co-author. He published more than 20 scientific articles and few extension manuals and booklets. He is life member in various fisheries societies and bodies. He serves as research advisory committee members for Post Graduate Students in his faculty. He handled more than 15 University funded projects as PI and Co-PI and organized three training programmes, one workshop, Webinar and one certificate course for the benefit of faculties and aqua farmers.

Dr. T.V. Ramana obtained his M.F.Sc., and Ph.D degree from College of Fisheries, Mangalore. He served in various capacities as Professor, University Head and Associate Dean and Dean of Fishery Science at Sri Venkatesawara Veterinary University, Tirupati. He was member of Peer Review Committee (PRT) of ICAR for accreditation of Fisheries Universities/Colleges. He served as one of the members in Andhra Pradesh state committee on constitution of fish feed and review of fish seed act. He served as expert for various fisheries colleges and Academic council and syndicate member for the discipline of Fisheries Biology and Resource Management in Tamil Nadu Fisheries University, Nagapattinam. He has been guiding master and PhD students in the discipline of Fisheries Resource Management.

He has published more than 25 research papers in National and International journals and few training manuals as author and co-author. He published more than 10 scientific articles and few extension manuals and booklets. He is life member in various fisheries societies and bodies. He serves as research advisory committee members for Post Graduate Students in the faculty. He handled ICAR and many University funded projects as PI and Co-PI.

Advance Trends in Aquaculture Practices

– Editors –

A. Balasubramanian

Professor and Head,
Department of Fisheries Resource Management,
College of Fishery Science,
Muthukur, Nellore District, Andhra Pradesh

T.V. Ramana

Principal Scientist and Head,
Fisheries Research Station,
Kakinada (Former Dean of Fishery Science, SVVU)

2023

Daya Publishing House®
A Division of
Astral International Pvt. Ltd.
New Delhi – 110 002

ISBN : 9789359191973 (Int. Edition)

Publisher's Note:

Every possible effort has been made to ensure that the information contained in this book is accurate at the time of going to press, and the publisher and author cannot accept responsibility for any errors or omissions, however caused. No responsibility for loss or damage occasioned to any person acting, or refraining from action, as a result of the material in this publication can be accepted by the editor, the publisher or the author. The Publisher is not associated with any product or vendor mentioned in the book. The contents of this work are intended to further general scientific research, understanding and discussion only. Readers should consult with a specialist where appropriate.

Every effort has been made to trace the owners of copyright material used in this book, if any. The author and the publisher will be grateful for any omission brought to their notice for acknowledgement in the future editions of the book.

Published by : **Daya Publishing House®**
A Division of
Astral International Pvt. Ltd.
– ISO 9001:2015 Certified Company –
4736/23, Ansari Road, Darya Ganj
New Delhi-110 002
Ph. 011-43549197, 8130496929
e-mail: info@astralint.com
Website: www.astralint.com

Acknowledgement

We acknowledge our profound gratitude to the Vice Chancellor of Sri Venkateswara Veterinary University, Tirupati for zealous encouragement for bringing out this book entitled *"Advance Trends in Aquaculture Practices"* for the benefit of the farmers, students, researchers and teachers in the faculty of fisheries and allied sciences.

Alike the editors are immensely grateful to all the contributors who shared their knowledge and experiences in their articles published in this book and other faculty members of College of Fishery Science, Muthukur for their constant encouragement at various stages of preparation of this book.

Editors are obliged to the NFDB and MANAGE, Hyderabad for permitting us to publish the lectures published in the training manual prepared in Aquaclincs and Aquapreneurship Development programme conducted during the year 2020.

We are obliged to extend our gratitude to all the authors and publishers whose books have been referred in preparing this book

In the last, but not the least, our publisher M/s. Astral International Pvt Ltd, New Delhi, deserves the highest gratefulness for instantly accepting our interest of bringing out this book under their scholasticism and publishing it well in time.

Prof. A. Balasubramanian

Prof. T.V. Ramana

Preface

Fisheries and Aquaculture are very important sector nowadays since they fetch a huge quantum of foreign exchange to the country. The increase of production of fish and fishery products purely relies on availability of knowledge, skill and new technologies. Unless introducing new technology and techniques in any production sector, the development cannot be achieved. In this juncture, introduction of new and advance techniques in aquaculture practices receives significant importance. Considering all these, the compendium titled "Advance Trends in Aquaculture Practices" was prepared by inviting articles in advanced techniques from various reputed scientists from national institutes like Indian Council of Agriculture Research and fisheries colleges of the country. Every author who contributed his article in this book has shared his knowledge and experiences acquired in his field of research and education.

Nonetheless many books are available in fisheries and aquaculture; it could be observed that very few books have been written with special emphasis on advance technologies in fisheries and aquaculture. In fulfilling such inadequacy, this compendium is compiled with articles imparting advanced knowledge on various diversified technology introduced in aquaculture in recent years like Recirculatory systems, Raceway system, Biofloc system, Cage farming, Aquaponics, Culture of Crab, Seabass culture, live feed culture, Indian Pompano and Grouper, *etc.*

Further, the compendium titled *"**Advance Trends in Aquaculture Practices**"* has been prepared intently for the benefit of the students in fisheries and other allied sciences as well as aqua farmers to update their knowledge in the field of aquaculture. Every chapter has been written meticulously in easy and simple language for making the readers to understand comprehensively of concepts and technology explained in every chapter.

It is confident that this compendium would be highly beneficial and trendy to all readers irrespective of the category. Besides, it would serve as handbook for the young researchers who involve in aquaculture research.

Suggestions and critique are hailed from the readers to make the treatise more useful to the readers in future.

Prof. A. Balasubramanian

Prof. T.V. Ramana

Contents

Chapter 1

Coastal and Brackish Water Aquaculture in India

V.S. Chandrasekaran

Principal Scientist (Retired),
ICAR-Central Institute of Brackish water Aquaculture, Chennai

Traditional capture fishery has reached maximum production levels and therefore it will not be able to provide estimated additional 37 million tonnes of fish needed to satisfy a growing world population by 2030. Currently 48 million tonnes of fish are destined to human consumption per year. It is estimated that in 20 years, the world's population will increase by two billion, and fish production will have to double in order to maintain current consumption levels.

Aquaculture is the highest ranking food sector with an annual growth rate of 8.8 per cent for the past 25 years. Livestock, for example, grew a mere 2.8 per cent during the same period. Currently, about 45 per cent of total fish allocated to human consumption comes from fish farms. Not only does aquaculture provide foods rich in proteins, fatty acids, vitamins and minerals, but also significantly aids in combating hunger and malnutrition. Moreover, the sector is known to contribute to a nation's economic development by increasing income and generating employment. In Asia the sector provides employment for 12 million persons.

Aquaculture in India

Aquaculture, though has more than 2500 years of history, its rapid growth has been experienced only since the last century. In India, although aquaculture has been practiced for centuries (as in China) the developments took place after attaining independence of the country, particularly in freshwater areas, and the development of more intensive brackish and coastal water aquaculture took off in the past few decades.

Out of a total 3.9 million ha of the brackish water area estimated, 1.2 million hectares of coastal saline lands have been identified to be potentially suitable for brackish water farming. Also, about 9 million hectares of salt-affected inland soils in the hot semi-arid and arid regions of northern plains and central highlands in the states of Haryana, Rajasthan, Punjab, Uttar Pradesh, Maharashtra and Gujarat are also found to be suitable. Estimates show that only 11 per cent of the potential coastal area available is utilized for aquaculture practices. Hence, there is a tremendous scope for development of coastal and brackish water aquaculture as high investment large scale farming system or as small area low or moderate economic small-scale aqua farming.

Coastal and Brackish Water Aquaculture in India

The extensive systems of traditional aqua farming mainly shrimp culture in West Bengal, Kerala, Karnataka and Goa are the main stay of brackishwater aquaculture in the recent past, contributing to the economic developments in the coastal belt of the country.

Mariculture includes (i) the coastal seawater based aquafarming such as, pen, cage, rack, raft and long line culture and (ii) offshore farming using floating and submersible cages which are sturdy enough to withstand rough weather and sea conditions in the high seas. In India, constant efforts have been made for the first category of aqua farming in the past five decades by the ICAR-CMFRI. Green mussel (*Perna viridis*) culture has been undertaken to some noteworthy extent in Malapuzham, Kasargod, Mangalore and Goa areas on the west coast of the country. Rope culture and long line culture techniques are used in deep water (5-10m depth) and shallow sea and backwater (<5m depth) areas, respectively for farming of edible oyster *Crassostrea madrasensis*,

In addition, the hatchery seed production and farming technologies have been developed by the ICAR-CMFRI for the species Silver Pompano, Cobia, Blue swimmer crab, *etc*. Captive broodstock development, induced spawning and seed production of cobia *Rachycentron canadum*, broodstock development, induced breeding, larviculture and seed production of silver pompano *Trachinotus blochii*, orange spotted grouper *Epinephelus coioides*, Indian pompano *Trachinotus mookalee*, captive breeding and seed production techniques of selected clown fishes and damsel fishes. Among crustaceans, the breeding and seed production technologies have been developed by the institute for the shrimp *Penaeus semisulcatus*, marine crab *Portunus pelagicus* and sand lobster *Thenus unimaculatus*.

Seaweed (*Kappaphycus alvarezii*) farming is gaining momentum, particularly with the coastal village communities and women Self Help Groups, in the Palk Bay and Gulf of Mannar regions and recently in the state of Kerala. Seaweed farming is practiced in the 5-10 m depth neritic zone immediately after the surf region from the sea shore. The women SHGs and others involved in seaweed farming faced problem in the forest reserve area of the neritic waters in Gulf of Mannar as it has been classified under the Man and Biosphere (MaB) and as National Park. However, the other coastal areas have free access for such activities and their technical suitability is to be assessed for seaweed culture.

Open water cage farming is an activity that will help to produce large amount of fish needed for our country. Potential sites, such as, sheltered bays in Ratnagiri, Goa, Karwar on the west coast and Palk Bay, Larsons Bay, Gulf of Mannar on the east coast and the Lakshadweep, Anadaman and Nicobar Islands, have been identified suitable for open sea cage culture. However, specific policy guidelines on leasing the sea areas for such mariculture activities, with due consideration for environmental and biodiversity conservation, are yet to be developed in our country, which would be essential for commercialization of these sectors. Pearl oyster culture (*Pinctada fucata*), spiny lobster culture (*Panulirus homarus*) and sea cucumber (*Holothuria scabra*) farming also attribute to the scope of diversification in mariculture in India. The recent developments in the sea cage farming and sustainable Capture Based Aquaculture (CAB) systems by the ICAR-CMFRI and Integrated Multi-trophic Aquaculture (IMTA) of Asian seabass and green mussels by the ICAR-CIBA are worth mentioning.

Diversification in Brackishwater Aquaculture

Crustaceans

Brackish water shrimp farming is a sector with a very high degree of diversity, involving a wide range of species, farming systems and production practices, and farming locations. There are significant differences between and within countries regarding the levels of production intensity and yields, farm numbers and their sizes, and the various types of resources utilized. The tiger shrimp *Penaeus monodon* was the predominant species for farming all over the world, because of many favourable traits, such as, fast growth, tolerance to wider environmental conditions, good meat quality, *etc.* Shrimp farms are often classified into extensive (low input systems characterized by low stocking densities, little or no external nutritional inputs, tidal water exchange and shrimp yields of less than 500 kg/ha/yr); semi-intensive (use of fertilizers combined with supplemental feeding, intermediate stocking, occasional pumping of water and yields of 1-2 t/ha/yr); and intensive systems (high stocking density, formulated complete feeds, aeration and water pumping with yields of more than 2 t/ha/yr). Such classification of shrimp farming systems is difficult, and can be rather arbitrary, given that there are additional characteristics and different criteria and terminologies in use.

Figure 1.1. *Penaeus monodon*

Farms may also use monoculture or polyculture systems (polyculture systems are usually common with low input systems); they may be operated as mixed systems (*e.g.* shrimp and mangrove farms); or by alternate cropping, involving one crop of shrimp followed by a harvest of another species or crop (*e.g.*, rice-shrimp alternate cropping systems in Bangladesh, India, and Vietnam). The size of farm is also very variable.

In Asia, small-scale farms dominate shrimp farming in many countries, which is in contrast to many farms in the Western hemisphere. In India about 90 per cent of the shrimp farms are classified as small farms, with a farm size of less than 2 ha. The use of terms such as 'traditional' and 'industrial' shrimp farming does not describe the extreme diversity of shrimp farming. Thus, an important consideration when discussing shrimp farming is the diversity of farming systems in operation as well as their location, size, management and the people involved. Presently low to medium stocked culture practices are being followed using tiger shrimp in India.

Western White Shrimp *Litopenaeus vannamei*

Dependence on the farming of single species, *viz.*, the tiger shrimp (*P. monodon*) in brackish water aquaculture sector has been reduced by the introduction of the western white shrimp *Litopenaeus vannamei* all over the world, replacing to an extent of about 90 percent of the tiger shrimp produced through aqua farming. However, in India after a thorough analysis of pros and cons at various levels and after making the risk assessment on its introduction, the western white shrimp *L. vannamei* has been permitted for culture in the year 2009. The *L. vannamei* has become the predominant species in the world's shrimp aquaculture production, including India, largely due to its low production costs, as low as half that of the black tiger production cost. The total shrimp, predominantly the Pacific white shrimp (*Penaeus vannamei*), produced through brackishwater aquaculture in India during the year 2017-18 was 6.2 lakh (6,22,327) tonnes, from 93,496 ha area, accounting to a productivity level of 6.6 t/ha/yr (MPEDA, 2019).

Figure 1.2. *Litopenaeus vannamei*

Importation, hatchery seed production and cultivation guidelines for the shrimp *L. vannamei* have been prepared by the Department of Animal Husbandry, Dairying and Fisheries, Government of India. The Coastal Aquaculture Authority (CAA) is the body that grants permission to import *L.vannamei* brood stock, its hatchery seed production. The farmers should also get themselves registered with CAA to carryout *L.vannamei culture.* The brood stock importation is being allowed only from a pre-approved group of suppliers. The supplier approval is being granted by the CAA in conjunction with the National Fisheries Development Board (NFDB), the Central Institute of Brackish water Aquaculture (CIBA) and the Marine Products Export Development Authority (MPEDA). The short listing of suppliers is being made on a genetic basis as well as on the disease status of the brood stock. According to the guidelines, all *L.vannamei* brood stock must enter the country through Chennai, wherein the Animal Quarantine Officer will inspect the consignment in question and send it to the quarantine facility when necessary.

Banana Shrimp *Fenneropenaeus merguiensis* and Kuruma Shrimp *Marsupenaeus japonicas*

Figure 1.3. ***Fenneropenaeus merguiensis***

The banana shrimp *Fenneropenaeus merguiensis* and the Kuruma shrimp *Marsupenaeus japonicus* are two alternative species for culture in India. The natural maturation of the banana shrimp has been achieved by CIBA in ponds and viable spawning has also been made subsequently. Natural maturation of kuruma shrimp *M. japonicus* was also achieved under controlled conditions in the hatchery by CIBA.

The seed production technology for the banana shrimp *F. merguiensis* has been developed in CIBA. The trial culture of banana shrimp was done in a farmer's pond under Public-Private Partnership mode at Billimora in Gujarat using the seed produced in CIBA.

Figure 1.4. ***Marsupenaeus japonicas***

The post larvae stocked @ 40,000 Nos./ha yielded a production of 903 kg/ha in 160 days. The shrimps attained final average size of 27 g with 80 per cent survival. The culture trials have confirmed the potential of banana shrimp as an alternate species for farming in winter period in Gujarat and Maharashtra when *P. monodon* exhibits slow growth in lower temperature period.

Culture of M. *japonicus* was also undertaken in a 0.05 ha pond at a stocking density of 10 No./m^2. In 154 days of culture, the shrimps attained average size of 13.3 g and yielded a production of 987 kg/ha. The kuruma shrimp fetches premium price as live material in Japan and other Southeast Asian countries like, Hong Kong, and has great potential for farming as an alternative species in India.

Indian White Shrimp *Fenneropenaeus indicus*

Figure 1.5. ***Fenneropenaeus indicus***

The Indian white shrimp (*F. indicus*) is found distributed from south and east African coasts, to India and Sri Lanka including Madagascar and the Red Sea. Further east it extends to south China, the Philippines and northern Australia. It inhabits shelf areas from the coastline to depths of about 90 m; most abundant in shallow waters in less than 30 m on sand or mud (slight preference for sandy bottoms); a very euryhaline species since in the Red Sea it tolerates very high salinities. In India it is one of the most abundant, and at the same time, is among the most highly priced shrimp; juveniles contribute to a good fishery in estuarine regions. *F. indicus* is an important species being cultured through semi-intensive and intensive culture systems in the West Asian Gulf countries and along the South East African region.

It has been cultured along with other shrimps and fishes in the extensive traditional culture systems in India. Monoculture of this species was carried out about two decades back in Tamil Nadu and a few other coastal states of the country. Since this species can withstand high salinities up to 45 ppt, it was a best suitable one for culture in summer in southern parts of the country, when the tiger shrimp could not withstand the high salinities. It has been also cultured in salt pan reservoirs in Tuticorin providing additional revenue to the salt producers.

At present, culture of this species is getting momentum again as it is preferred species for domestic marketing in India. Because of the recent problems in tiger shrimp farming, such as low international competitive price, stringent import regulations like traceability, high production cost and low profit margin, culture of Indian white shrimp has been found to be a better candidate. The recent trend of *F.indicus* culture, comprised of stocking hatchery produced quality seeds at moderate stocking densities around 10 PL/m^2, culturing up to 75 DOC, subsequently periodical harvesting in small quantities to meet the local market demand on weekly basis, up to 120 days *etc.*, has been found to be a sustainable venture by the small farmers.

Mud Crabs *Scylla serrata*

Figure 1.6. *Scylla serrata*

Mud crab farming is an important alternative source of income for aqua-farmers in India. On an average 1,500 tonnes of live mud crabs are being exported from India that fetches an earning of Rs. 26 crores per year. This includes the crabs captured from the wild as well as the ones cultured/ fattened. The expanding export market for mud crab as an alternative for shrimp has led to intensified collection of wild seed for grow-out and fattening and has threatened the wild stocks. To ensure the sustainability of crab farming and reduce the fishing pressure on wild stocks, CIBA has developed the hatchery technology for the production of seeds of mud crab, *Scylla serrata* and also the nursery rearing technologies. The mud-crab hatchery technology developed at CIBA has also been transferred through training to a private entrepreuner who has established a commercial scale mud crab hatchery, the first of its kind in the country, at Mugaiyur, Chengalpet taluk, Tamil Nadu.

Mud crab farming can be classified into (i) the long term grow-out culture and (ii) the short term fattening method. Fattening of mud crab is done either to (i) gain weight in moulted in 3 to 4 weeks or (ii) to attain full development of ovary in adults in 3 to 4 weeks. The mud crabs can be cultured in three types of culture systems in brackishwater areas, *viz.*,1. Fenced earthen ponds, 2. pens in open backwaters, 3. Cages. The technology of mud crab fattening in pens and cages are a sort of small scale farming enterprise suitable for coastal fisher folk as an additional means of income generation. It is also quite suitable for adoption by coastal women, as a small-scale enterprise, for their economic upliftment. The indoor vertical rack system for mud crab fattening is in vogue in the south east Asian countries like in Malaysia, Singapore, Thailand, Vietnam, *etc.* This technology has a scope for expansion in India also with the incorporation of a safe healthy RAS system.

Finfishes

Asian Seabass *Lates calcarifer*

Figure 1.7. *Lates calcarifer*

Asian seabass (*Lates calcarifer*) known as 'Bhetki' in many parts of India is an important food fish. Though it forms negligible in perch fishery in commercial catches, it is highly suitable for farming. Seabass is a euryhaline species capable of withstanding wide salinity fluctuations. It is a fast growing fish and suitable for culture both in earthen ponds and in cage system. Amongst the cultivable fishes in India, seabass fetches higher price in domestic market varying between Rs.200-450 per kg depending upon the size, the availability and season. It is extensively cultured, in South East Asian Countries like Thailand, Malaysia, Singapore and Australia. In India, culture of seabass in a limited level is carried out in the traditional farms in coastal areas and on semi-intensive scale in a few pockets of coastal brackish water bodies in states like Andhra Pradesh, Orissa, West Bengal and Maharashtra. Compared to the carp culture in fresh water or shrimp culture in brackish water ponds, seabass culture is still in its infant stage in India. The main reason for this is that the market for shrimp is global while the market for seabass is largely regional. However seabass is the only species, most commonly farmed amongst fin fishes in South East Asian countries because the seed can be produced in the hatchery economically throughout the year at reasonable price.

Trial culture of seabass conducted at CIBA adopting polyculture method yielded around 4.0 tons/ha for 11 months culture duration. Hatchery produced seabass fry of 1.0 cm were reared in hapa nursery at the culture pond site. After one month rearing, the fry attained an average size of 1g. In a 1.0 ha area pond 8,000 seed of seabass were stocked and reared under polyculture method, along with Tilapia. At the time of harvesting the average size of the fish was 740 g (ranging from 300 to 1500 g) and the recovery rate was 68 per cent. The total expenditure incurred was Rs.62,300 and the realization was Rs.2,37,000 with a profit of Rs.1,74,700/- under the polyculture method.

The economics worked out from the field trials conducted in the culture of Asian seabass feeding with low cost fishes are as follows: From the commercial landings, low cost fishes like oil-sardine, horse mackerel *etc.*, were procured @ Rs.3-6/kg as used as feed for seabass. The pond was stocked @5000/ha of average size 1.0 g. After 11 moths culture period the fish grew to size of 517 to 1200 g (average size of 946.5 g) with the recovery rate of 96 per cent. The production was in the order of

4.36 tons/ha. The realization was Rs. 3,61,585/- with expenditure of Rs. 3,14,800/-. The profit over the operational cost was to the tune of Rs. 46,785/-

Under cage culture, since seabass can be intensively stocked and properly managed, the production will be high. Frequently culling and maintenance of uniform sized fishes in to the cages will ensure uniform growth and high production. Production of 6-8 kg/m^2 is possible in the cages, under normal maintenance and production as high as 20-25 kg/m^2 is obtained in intensive cage management in the culture of seabass.

CIBA has achieved a major breakthrough in the captive broodstock development, induced breeding and seed production of the Asian seabass *Lates calcarifer in 1997* for the first time in the country. The technology for seabass seed production has been also standardized. The existing facility of the Institute for controlled breeding and seed production has been expanded by the addition of a hatchery under Indo-French collaborative project "Seabass breeding and culture – pilot unit" with one million fry production capacity. This additional facility is the first model finfish hatchery in the country incorporating all the required facilities with biosecurity and water recirculation system for quality seed production. It is possible now to produce the seabass seeds round the year and even in low salinities of around 15 ppt.

CIBA has been conducting field demonstrations and trail culture of seabass in farms in the east and west coasts of the country, *i.e.*, Tamil Nadu, Andhra Pradesh, Maharashtra and Gujarat in order to develop the culture technology packages for the different coastal agro-climatic conditions of our country. The nursery rearing technology has also been developed by CIBA and implemented as a small scale operation at field level. The low cost pellet feed "Bhetki ahaar" formulated specifically for the seabass in CIBA has also been found to be efficient and is being used for field trial culture to confirm its performance.

Mullet *Mugil cephalus*

Figure 1.8. *Mugil cephalus*

The mullets are one of the important table fishes in the Indian market and has occupied the major catch of fishes in the traditional brackishwater polyculture systems like *'Bhasabadha'* and *'Berri'* in West Bengal, *'Pokkali'* fields of Kerala and other traditional aquaculture practices in Goa, Karnataka and other coastal locations of the country. Monoculture of mullets on scientific basis holds promising future as mullets are now a days fetching premium price in international markets. However, seed availability at required quantity and season are the constraints. Full-scale commercial production of *Mugil cephalus* is not yet common. Induced spawning and production of fry has been achieved on an experimental and semi-commercial basis in the United States of America and Taiwan Province of China, and the production of mullet fry on a limited scale for aquaculture has been reported in Italy, Israel and Egypt. CIBA is in the processes of developing hatchery technology for the production of the Grey mullet *Mugil cephalus* seeds and also gold spot mullet *Liza parsia*.

Milk Fish Culture *Chanos chanos*

Figure 1.9. *Chanos chanos*

Milkfish is a warm water species and prefers water temperatures between 20-33°C. Unlike many other large saltwater fish it is herbivorous and feeds on diatoms and other similar food items. The broodstock can be raised and spawned in captivity to produce larvae in the hatchery. It spawns only in fully saline water and releases up to 7 million epipelagic eggs (1.1-1.2 mm diameter) which hatch in about 24 hr. In nature, their larvae seek out clear coastal and estuarine waters warmer than 23ºC with 10-32 ppt salinity and abundant phytoplankton. Incubating eggs and newly hatched larvae are transported to the shore by currents, where they are gathered by people in brackish

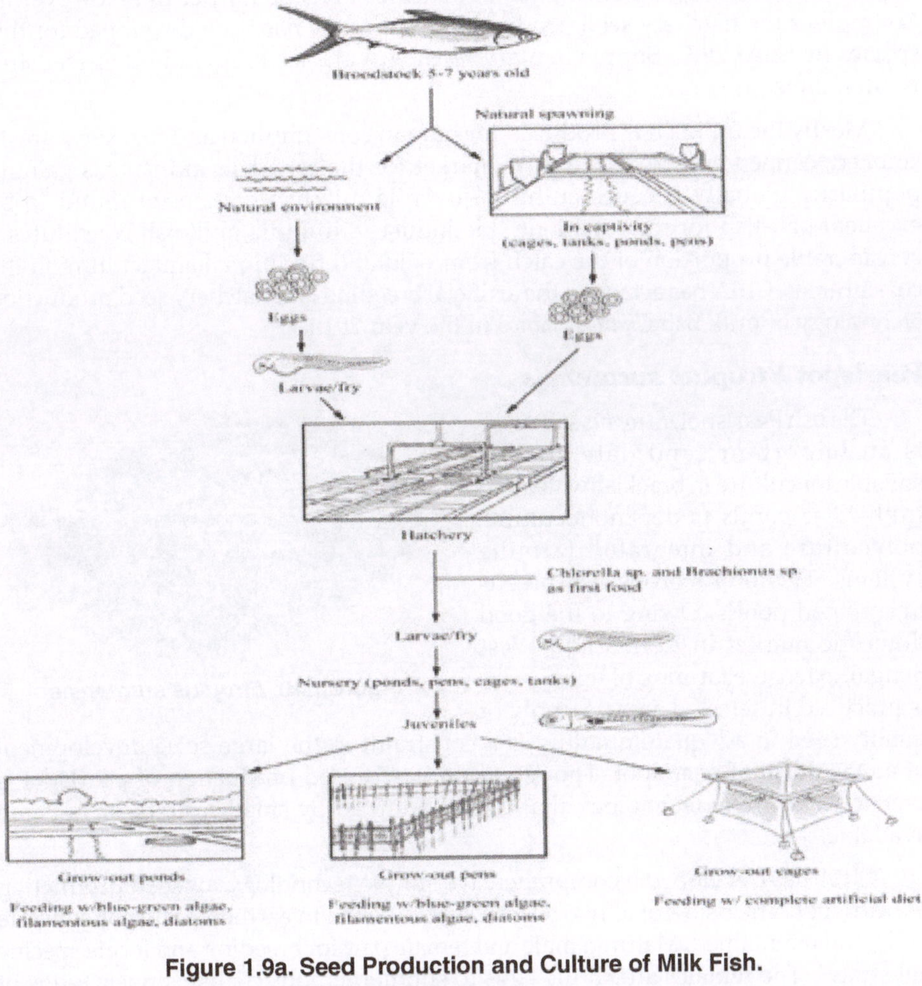

Figure 1.9a. Seed Production and Culture of Milk Fish.

waters such as shallow sandy areas, mouths of rivers, and lagoons. Older larvae migrate onshore and settle in coastal wetlands (mangroves, estuaries) during the juvenile stage, or occasionally enter freshwater lakes. Under natural conditions, larvae and fry migrate inland seeking tidal pools and settle in them for 1 month until they become juveniles, then migrate into lagoons, lakes and shallow waters until they reach adolescence (24-45 cm fork length) before returning to the sea for further grown and sexual maturation. Milkfish attains sexual maturity in 5 years, both in the wild and under culture conditions.

Culture of this species is well developed in South East Asian countries, especially the Philippines, Indonesia, Taiwan and Thailand. Larvae are collected from rivers and are grown in culture ponds into juveniles. Since milk fish is herbivorous, culture is by the extensive method, where, large bodies of pond waters over 1 to 5 hectares are fertilized with organic and/or inorganic fertilizers to enhance primary production. Fry are caught from the shallow tidal pools and lagoons and stocked in pre-fertilized culture ponds often at densities of 30-50 fry per m^2. However, in Philippines the hatchery seed production technology has been developed for this species by SEAFDEC. Supplementary feeding with rice bran and other feedstuff is often done.

Mostly the milkfish is produced for human consumption and marketed fresh, smoked, canned and frozen. A new market for the tuna bait industry is gaining popularity globally because cultured juvenile milkfish have been found to be excellent bait-fish for the tuna long line industry. In India, milkfish constitutes a considerable proportion of the catch from wild and traditional aquaculture in the coastal areas. CIBA has achieved the artificial breeding and hatchery seed production technology of milk fish *Chanos chanos* in the year 2015.

Pearlspot *Etroplus suratensis*

The fish Pearlspot, *Etroplus suratensis* is an important candidate species suitable for culture in brackishwater and freshwater ponds under monoculture, polyculture and integrated farming systems. This omnivorous fish breeds in confined ponds. Owing to the good domestic market in Kerala and West Bengal, extensive farming of this species is practiced in selected areas. Supply of quality seed in adequate quantity is a constraint in the large scale development of aquaculture of pearlspot. Though pond based seed production of pearlspot is practiced, a simple technology that can be adopted by small scale farmers is not available.

Figure 1.10. *Etroplus suratensis*

CIBA has developed a comprehensive simple technology for seed production of pearlspot. The fishes of above 50 g size are stocked in earthen ponds or concrete tanks and maintained. Matured male and female pair for breeding and locate specific substrates. The females attach the eggs to submerged objects like stones, twigs of wood or any flat hard object. The fertilized eggs undergo embryonic development and hatch out after 6-7 days. Just prior to hatching, the eggs are transferred to pits (nest) made by the parent fishes. The parents guard the juveniles and guide for feeding and offer care till they reach 4-5 cm. From a single pair, 600-3000 juveniles can be obtained. This technology has been developed for breeding of pearlspot for longer period in a year in confined ponds and tanks by manipulating the environmental conditions such as salinity around 10-15 ppt.

Cobia *Rachycentron canadum*

The marine fish Cobia, *Rachycentron canadum*, is a widely distributed species from the Indo-Pacific waters to the southern Atlantic Ocean. It is an ideal species for nearshore as well as land based cage culture. Due to its high market value in both domestic and international markets, the technology for its culture has rapidly developed in the past few years globally. Grow-out production of cobia in cages is ongoing or has occurred in China, particularly Taiwan, USA, the Bahamas, Belize, Dominican Republic, Mexico, Puerto Rico, Philippines and Viet Nam with additional projects reportedly planned for the Caribbean and Central America; however, there was not any large scale commercial production of cobia reported in 2006 and the industry in the western hemisphere could be characterized as 'developing'.

Figure 1.11. ***Rachycentron canadum***

Production Cycle of *Rachycentron canadum* (FAO)

The technologies developed in various countries include mass propagation through natural spawning of captive broodstocks, larval rearing techniques, nursery production in tanks, ponds and nearshore cages, and grow-out culture in offshore cages. Reproduction in captivity is relatively easy because sexual maturity often occurs within 2 years of culture. In Taiwan it has been reported that spontaneous spawning occurs year around at water temperatures of 23–27°C, with peak during spring and autumn. The fertilized eggs hatch within 21–37 h at water temperature of 31–22 °C. The larval growth is fast, and larvae are vitally robust and environment tolerant compared to other marine fishes. The fry can be mass-produced in outdoor ponds at relatively low cost. Weaning of fry from 20 days old onwards to pellet feeds is feasible. Nursery rearing from 10–30 to 1000 g can be done in either outdoor ponds or nearshore cages.

In recent years, intensive and super-intensive recirculation systems for nursery (from 2 to 100–150g) were developed with survival rates of more than 90 per cent. In nursery and grow-out offshore cages, 100 –600 g cobia were cultured within 1–1.5 years when they reached 6–8 kg for export to Japan, or 8–10 kg for the domestic market. Currently, around 80 per cent of marine cages in Taiwan are devoted to cobia culture. However, some problems still exist in cobia culture that needs to be addressed and solved to increase production. These include high mortality due to stress during transport from nursery tanks/inshore cages to grow-out cages and diseases during nursery and grow-out culture resulting in low survival, and

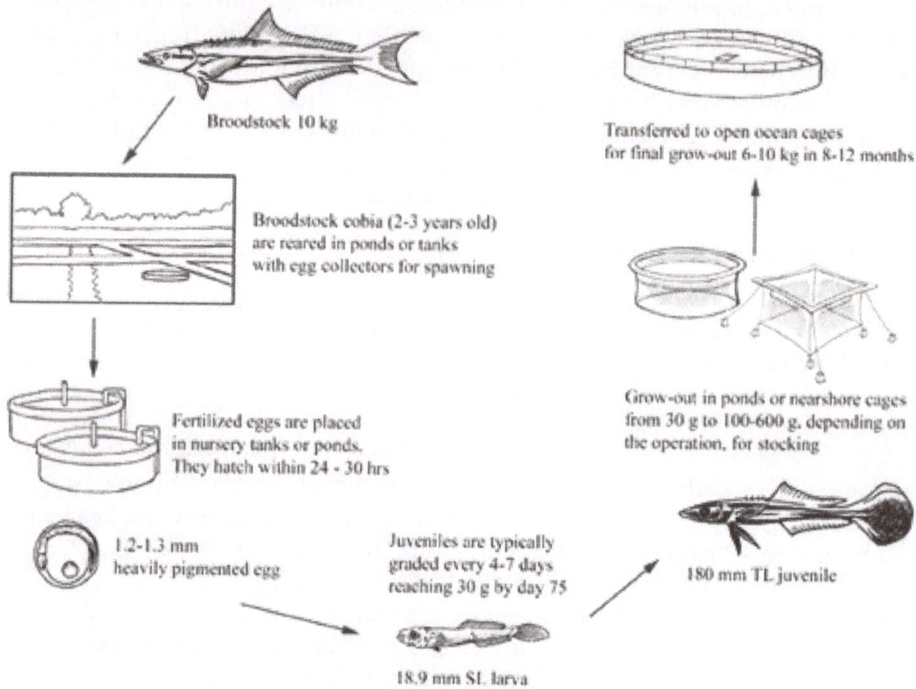

Figure 1.11a. Production Cycle of *R. canadum*.

consequently poor harvest. In India, the technologies involved in Cobia culture has been developed by the RGCA and propagated for commercial operations.

Brackishwater Ornamental Fish Culture

Ornamental fishes from brackishwater ecosystem are not being exported from India commercially because of the limited resource and sporadic distribution. However, it can be foreseen, if the large scale commercial production of the brackishwater ornamental fishes *in par* with the freshwater ornamental fishes, there is a great scope for this sector in our country.

CIBA has initiated the development of captive broodstock and breeding techniques for the brackishwater ornamental fishes *Scatophagus argus* and *Monodactylus argenteus* under controlled conditions and larval rearing techniques.

Spotted Scat *Scatophagus argus*

Figure 1.12. *Scatophagus argus*

The Spotted Scat is found distributed in India in most of the places along the east and west coasts of India and occur in abundance in some brackishwater estuarine systems. They are omnivorous. The juveniles appear round in shape with variable colours but basically a spotted brown to green. Red

Scat and Green Scat will turn to brown as they grow to 5 inches. The fish reaches 30 cm in length. Breeding of this species has been rarely reported. The young ones of scat are exported to USA from Asia for aquarium purpose and the cost is Rs 50-75/- per piece in India.

Monos *Monodactylus argenteus*

Figure 1.13. ***Monodactylus argenteus***

Monos (*monodactylus argenteus*) are disc shaped fish from the coastal areas of Australia and the Red Sea. They are often called as finger fish. They belong to the family Monodactylidae. In India they are found distributed in Gulf of Mannar. They have been reported from other coastal waters, estuaries, rivers, and lagoons in fresh, brackish, and sea waters. However, they are ideally suitable for brackishwater aquarium They are omnivorous and found to feed on flake, pellets,etc. The colour pattern of the young fish often have a black line running across the gill cover connecting with the black of the anal fin and the dorsal fin is marked with yellow tinge. The sexes are separate in this fish but difficult to distinguish. Successful captive breeding has not yet been reported for this species. On aquarium commercial trade this fish has, very good demand, costing 4 - 5 Singapore Dollar/ piece, and the local price in India is Rs 75-100/piece of 4 inches size. This would be an important potential species for aquaculture in brackishwater areas of India.

Other Finfishes

The other finfishes which are in the process of development of induced breeding, hatchery seed production and grow out farming for ornamental and/or food purpose are the streaked rabbit fish *siganus javus*, orange chromide *Etroplus maculates*, brackishwater catfish *Mystus gulio*, Goldspot mullet *Liza parsia* domestication and development of captive brood stock of Hilsa *Tenualosa ilisha*, broodstock development of canara pearlspot *Etroplus canarensis*, potential brackishwater ornamental fish species from Sundarban area, *viz.*, green pufferfish *Dichotomyctere fluviatilis*, mudskipper, *Pseudapocryptes elongatus* and Boddart's goggle-eyed goby/ mudskipper *Boleopthalmus boddarti*.

Entrepreneurship Development in Coastal and Brackishwater Aquaculture

The coastal and brackish water aquaculture sectors offer a wider scope for entrepreneurial options both at large scale and small scale levels. The high investment operations are mostly doing commercial shrimp monoculture, biofloc shrimp or fish farming, high density culture of fishes/shrimps in RAS facilities and large scale shrimp hatchery establishments. The low or moderate investment farming systems include nursery rearing of shrimp or fish fattening of mud crabs, running the production units and retail outlets of ornamental fishes, allied sectors or aqua feed production addressing specifically the commercially important shrimps, fishes, crabs and other organisms and allied sectors like the trade in farm inputs, establishment of health care units like aquaclinics, *etc.*, as per the requirements both at time and geographical locations. Whatever the technologies or species are going to be involved with as an economic activity the following would be some important steps that are to be focused for systematic progress:

1. Production technology/technical component of the enterprise
2. Scope of the enterprise/activity
3. Profitable scale of operation
4. Start up requirements– Land/farm/pond/machinery/men and material
5. Critical inputs, availability, sources and other considerations
6. Current level of penetration as a commercial entrepreneurial activity (Livelihood/Intermediate/Corporate)
7. Actual/potential clients of the enterprise/activity
8. Value chain and actual/potential market
9. Risks associated with the enterprise/activity and measures to overcome
10. Institutional support available for the activity/enterprise (technology, credit, subsidy, schemes)
11. One case/success story of the enterprise/activity
12. Regulatory guidelines and prescribed quality standards
13. Economics: Investments, Expenditures and Returns on investment
14. Important contact addresses pertaining to the enterprise/activity

Unlike fresh water fish culture, the brackish and coastal water aquaculture sectors are quite dynamic and continue to grow in terms of technology developments as per the diversified and new candidate species in demand in India. This also requires a lot of inputs, including strong policy guidelines. The Coastal Aquaculture Authority and the various Government R and D agencies and academic institutes have been making great efforts in formulating 'Good Management Practices' in the farming of shrimp, fish and other organisms in the coastal regions of the country, keeping in view of the welfare of the people, their environment and for the sustainable development of coastal and brackishwater aqua farming sectors, in perpetuity.

References

1. Rodrigues, J.G. and Kruse, M. 2017. Marine and coastal cultural ecosystem services: knowledge gaps and research priorities. One Ecosystem, 2.
2. Cheran, P.E (2018) Shrimp Aquaculture – Industry review, Aquaculture Spectrum Vol.1, issues 1 to11. Aquaculture Outlook, Chennai.
3. FAO (2018) The state of world fisheries and aquaculture (2016) - Contributing to food security and nutrition for all. Food and Agriculture Organization, Rome.
4. FAO (2016) National aquaculture sector overview India. http://www.fao.org/fishery/countrysector/naso_india/en/(Accessed on 07.06.2017).

5. Brugére Cecile, José Aguilar-Manjarrez, Malcolm C. M. Beveridge and Doris Soto (2018), The ecosystem approach to aquaculture 10 years on – a critical review and consideration of its future role in blue growth, Reviews in Aquaculture, p. 1–22.

6. Ayyappan, S.(2014) National Aquaculture Sector Overview. India. National Aquaculture Sector Overview Fact Sheets. In: FAO Fisheries and Aquaculture Department [online]. Rome. Updated 4 April 2014.

7. Krishnan, M., Birthal, P.S., Ponnusamy, K., Kumaran, M. and Singh, H. (2000) Aquaculture in India: retrospect and prospects. In: Aquaculture Development in India (eds M. Krishnan and P.S. Birthal), pp 11-31, Workshop Proceedings 7, National Center for Agricultural Economics and Policy Research, New Delhi – 110 012.

8. Marine Products Export Development Annual Report, 2017-18.

9. Mondal A, Chakravortty D, Mandal S, Bhattacharyya SB, Mitra A (2015) Feeding Ecology and Prey Preference of Grey Mullet, Mugil cephalus (Linnaeus, 1758) in Extensive Brackish Water Farming System. J Marine Sci Res Dev 6: 178. doi:10.4172/2155-9910.1000178.

Chapter 2

Aqua Farming Systems in India

D. Ravindra Kumar Reddy and R. Ram Mohan

Department of Aquaculture, College of Fishery Science, S.V. Veterinary University, Muthukur, Nellore District, A.P.

Indian fisheries and aquaculture is an important sector of food production providing nutritional security, besides livelihood support and gainful employment to more than 14 million people, and contributing a major chunk to agricultural exports. With diverse resources ranging from deep seas to lakes in the mountains and more than 10 per cent of the global biodiversity in terms of fish and shellfish species, the country has shown continuous and sustained increments in fish production since independence. The total fish production during 2017-18 is estimated to be 12.60 million metric tonnes. Of which nearly 65 per cent is from inland sector and about 50 per cent of the total production is from culture fisheries, and constitutes about 6.3 per cent of the global fish production.

Status of Indian Fisheries - 2017-18

Global position	3rd in Fisheries and 2nd in Aquaculture
Contribution of Fisheries to GDP (per cent)	0.91
Contribution to Agril. GDP (per cent)	5.23
Per capita fish availability (Kg.)	9.0
Annual Export earnings (Rs. In Crore)	45,106.89
Employment in sector (million)	14.0

Source: NFDB, GOI.

Cultivable Fish Species in India

- ✰ While carps form the most important species farmed in freshwater in India, it is the shrimp from the brackish water sector which contributes the bulk of the production. The three Indian major carps, namely, catla (*Catla catla*), rohu labeo (*Labeo rohita*) and mrigal carp (*Cirrhinus mrigala*) contribute over 90 per cent of the total Indian aquaculture production.
- ✰ The carp polyculture system was introducted during the year 1970 in the country with three exotic carps, namely, silver carp (*Hypophthalmichthys molitrix*), grass carp (*Ctenopharyngodon idellus*) and common carp (*Cyprinus carpio*). Among the catfishes, Philippine catfish, 'magur' (*Clarias batrachus*) which is the only species that has received a lot of attention and stinging catfish, 'Singhi' (*Heteropneustes fossilis*), another air-breathing catfish species are being cultured to a certain extent in swamps and derelict water bodies, especially in the eastern states.
- ✰ In recent years, attempts have been made to develop the culture of fishes like *Pangasius pangasius, Pangasius sutchi, Oreochromis niloticus, Ompok pabda etc.* The other finfish species of importance include climbing perch (*Anabas testudineus*), murrels (*Channa striata* and C. *marulius*), *etc.* Among the freshwater prawns, the giant river prawn (*Macrobrachium rosenbergii*), is the most important species followed by the monsoon river prawn, *M.malcolmsonii*.

Advanced Aqua Farming Systems in India

Aquaculture must have to move towards intensification to meet the rising demand, to contribute more effectively for reduction of poverty and malnutrition, and to become ecologically more sustainable. New technologies will make it possible for sustainable aquaculture to become new global standard. In order to improve the socio-economic condition of the farmers, this expansion of aquaculture production needs to take place in a sustainable way through the applications of new farming interventions *viz.* integrated farming, Aquaponics, Recirculatory Aquaculture System (RAS), Neo-female Technology, Biofloc technology (BFT), Compensatory growth Technology *etc.*

Unless the aquaculture industry is able to boost productivity, the limited availability of land, water, and feed may constrain its growth. The present aquaculture systems faces many challenges, mainly, water-quality management, harmful diseases and epizootics, development of appropriate feeds and feeding practices, hatchery as well as grow-out technologies.

1. Biofloc Technology (BFT)

Biofloc technology is a technique of enhancing water quality in aquaculture through balancing carbon and nitrogen in the system (Crab *et al.*, 2012). It is the retention of waste and its conversion to biofloc as a natural food within the culture system (Avnimelech, 2009). Bioflocs are the aggregates (flocs) of algae, bacteria, protozoans, and other kinds of particulate organic matter such as faeces and

Figure 2.1. Biofloc Technology (BFT).

uneaten feed. Each floc is held together in a loose matrix of mucus that is secreted by bacteria. In this system, the addition of carbohydrates to the pond, heterotrophic bacterial growth is stimulated and nitrogen uptake through the production of microbial proteins takes place (Avnimelech, 2009). If carbon and nitrogen are well balanced in the solution, ammonium with organic nitrogenous waste excreted from fish will be converted by the bacterial biomass. This system requires the addition of extra carbon to the aquaculture system, through an external carbon source or elevated carbon content of the feed. This system promoted nitrogen uptake or Immobilization of ammonium by heterotrophic bacterial growth decreases the

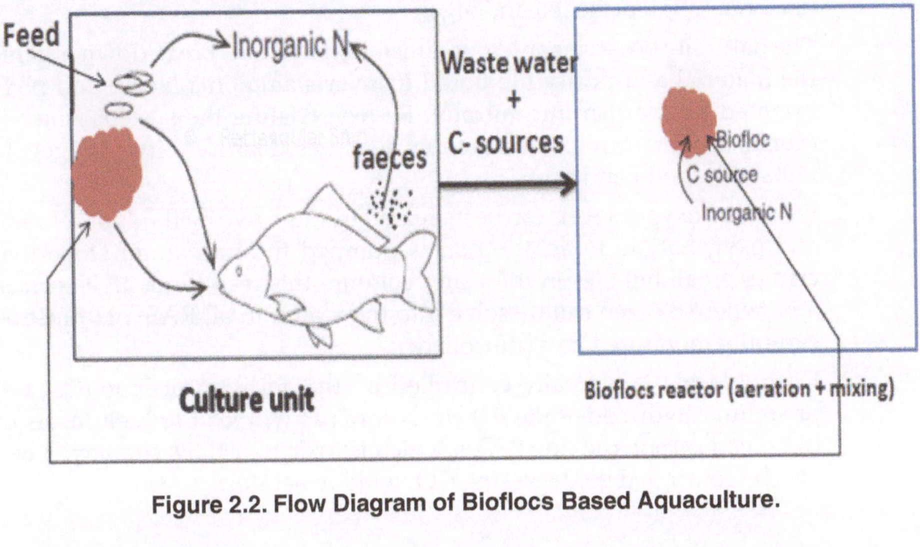

Figure 2.2. Flow Diagram of Bioflocs Based Aquaculture.

ammonium concentration more rapidly than nitrification. Thus, the BFT provides a new farming approach to increase the food production in a sustainable ways.

Advantages

- Bioflocs as a feed for aquaculture species.
- Bioflocs as a bio-control measure.
- Maintaining water quality by reducing toxic ammonia.
- Minimum or zero water exchange.
- Reducing feed conversion ratio and a decrease of feed costs.
- Nutrients could be continuously recycled and reused.
- Control pathogens in aquaculture.

2. Recirculation Aquaculture System (RAS) Technology

Recirculation aquaculture system (RAS) technology is the land-based closed systems in which aquatic organisms are cultured through the minimal use of water which is serially reconditioned. This land-based closed containment system improves food security and reduces environmental impacts.

- RAS consists of a series treatment processes removes organic and other oxygen demanding materials such as suspended solids, nutrients, fats, oil and pathogens from the waste water so that the water can be safely reused. The larger solid suspended solids, debris and floating materials (wood, paper rags and plastics) are removed by passing the waste through Mechanical filers *viz.* settlement tanks, sand filter, drum filter, screen filter, *etc.* where most of the solid materials removed.
- Then it is treated through a bio filters where the liquid waste is passed through a film of microbial growth develops on the filtering medium (Zoological film) which may be of 0.2-2.0 mm thick. The film consists of bacteria, fungi, protozoa and algae.
- The naturally occurring microorganisms on the films break down organic the material and purify the liquid (converts ammonia NH4+ and NH3 excreted by the fish into nitrate). Re-oxygenating the system water is crucial to obtain high production densities. Fish requires oxygen to metabolize food and grow.
- Dissolved oxygen levels can be increased through two methods of aeration and oxygenation. In aeration air is pumped through an air stone that creates small bubbles in the water column, this result in a high surface area where oxygen can dissolve into the water. In all RAS, pH must be carefully monitored and controlled.
- Desirable pH is typically controlled by the addition of lime ($CaCO_3$) or sodium hydroxide (NaOH) *etc.* A low pH will lead to high levels of dissolved carbon dioxide (CO_2), which is toxic to fish. Desirable pH can also be controlled by degassing CO_2 with an aerator.

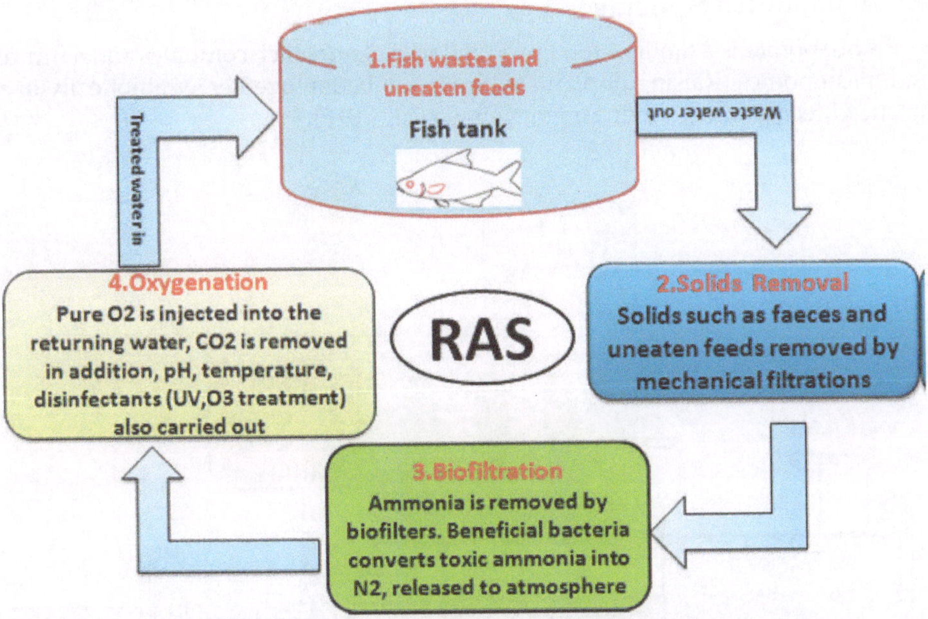

Figure 2.3. Flow Diagram of Recirculation Aquaculture System.

- ☆ All fish species have a preferred temperature above and below which fish will experience negative health effects and eventually death. Temperature also plays an important role in dissolved oxygen (DO) concentrations so; it can be controlled through the use of submerged heaters or heat pumps or chillers or heat exchangers. Finally it is disinfected using UV radiation or Ozone treatments before reused in the culture units.

Advantages

- ☆ Low water requirements.
- ☆ Low land requirements.
- ☆ Water quality parameters can be easily rectified.
- ☆ Independence of adverse weather conditions.
- ☆ High stocking density of desired species and productions.
- ☆ Most efficient; feed conversion is near 1:1.
- ☆ Reduce or eliminate vaccine, antibiotic and pesticide uses.
- ☆ Consistent production.
- ☆ Eco-friendly.
- ☆ Improve health and performance of the fish species

3. Aquaponics Systems

Aquaponics is a modern food production system which combines aquaculture and hydroponics (Raising of plants without soil beds) together symbiotically in a balanced recirculatory environment (Azad *et al.*, 2016).

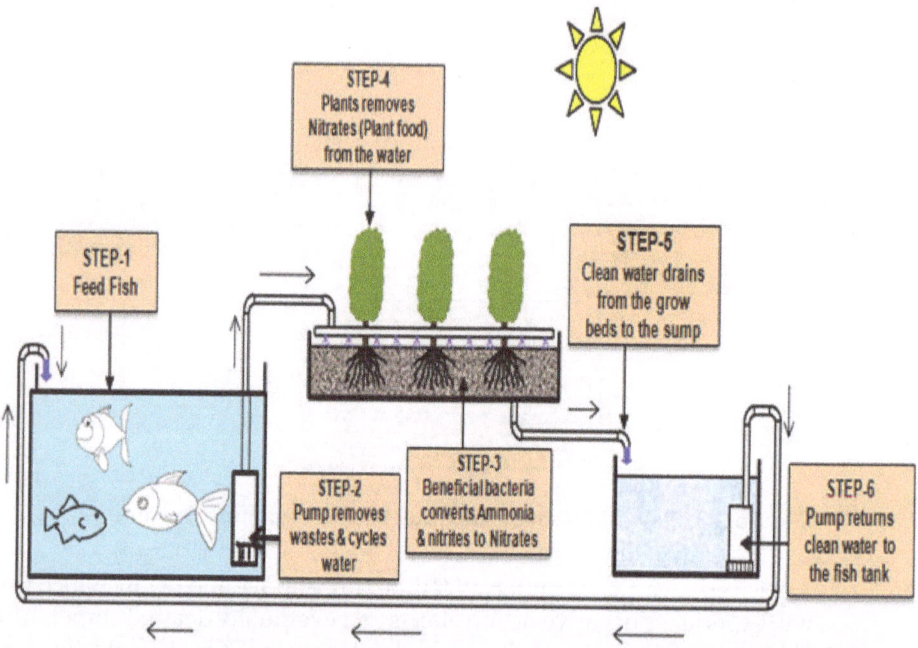

Figure 2.4. Flow Diagrammmatic Representation of Aquaponics System.

In Aquaponics system, nutrient-rich water from fish tanks is used as liquid fertilizer to fertilise hydroponic production beds. These nutrients in the water produced from fish manure, algae, and decomposing fish feed which otherwise increases the toxic levels in the fish tanks affecting the fish growth. The hydroponic beds function as a bio filter stripping off ammonia, nitrates, nitrites, and phosphorus so the freshly cleansed water can then be recirculated back into the fish tanks.

- ✰ The nitrifying bacteria living in the gravel and in association with the plant roots play a critical role in nutrient cycling. These nitrifying bacteria convert ammonia to nitrate, a form of nitrogen utilised by the plants. Thus when the water returns to the fish tanks, nitrogen level are tolerable for the fish. Unlike that of traditional farming, in aquaponics system there is a constant flow of water and constant supply of nutrients to the plants occurs.

Advantages

- ✰ Significant reduction in the uses of water
- ✰ No need to use artificial fertilizer

- ✰ Provide an artificial filtration system of fish culture environment.
- ✰ Does not require farmland with fertile soil.
- ✰ Naturally organic
- ✰ Tourist attractors in rural communities
- ✰ Reduced damage from pests and disease
- ✰ No weeding
- ✰ Plants grow faster
- ✰ Sustainable
- ✰ Cleaner form of gardening
- ✰ Easy to setup
- ✰ Eco-friendly

4. Mono Sex Culture or Neo-female Technology

Mono-sex culture is a farming practice based on the culture of fish by producing all males or all females' population depending upon the sex which have better food conversion ratio and growth rate. Generally monosex culture of all female population of Carp, Salmon and all male population of Giant freshwater prawn and *Tilapia* is carried out that maximize the production level. Among freshwater shell fishes culture *Macrobrachium rosenbergii* (Scampi) is gained importance in India. In this case the males reach market size faster than females. Thus, an all-male monosex population culture of the species is desirable. So monosex culture can be carried out by identifying the correct sexual dimorphism of this species and subsequently culture and harvest when attains desirable marketable size.

- ✰ Now-a-days a standardize technology called "neo female technology" where females are obtained through the sex reversal of males and that yield all male progeny. In this techniques the sex of the juvenile males are changed through microsurgical removal of the androgenic gland (AG) or through androgenic gene silencing (RNA interference method) to female (termed "neo-females"- phenotypic females with male genotype) and when it mate with a normal male gives all male progenies (Amir, 2013).
- ✰ In India, this neo-female technology project for the production of male scampi seeds has been undertaken by RGCA, Tamil Nadu and supplies all-male scampi seeds to farmers of the country.

Advantages

- ✰ Increase production to 2-flods.
- ✰ Better food conversion ratio
- ✰ Better growth rate

5. Integrated Fish Farming (IFF)

Integrated farming is defined as the sequential linkages between two or more agri-related farming activities with one farming as major component. When fish becomes the major commodity in the system, it is termed as integrated farming (Ayyappan, 2011). OR, integrated fish farming systems refer to the production, integrated management and comprehensive use of aquaculture, agriculture and livestock, with an emphasis on aquaculture. The linkage of fish farming with agriculture and animal husbandry is considered as sustainable farming system, which offers greater efficiency in resources utilisation, reduces the risk by diversifying crops, providing income and increasing food fish production for small scale farming.

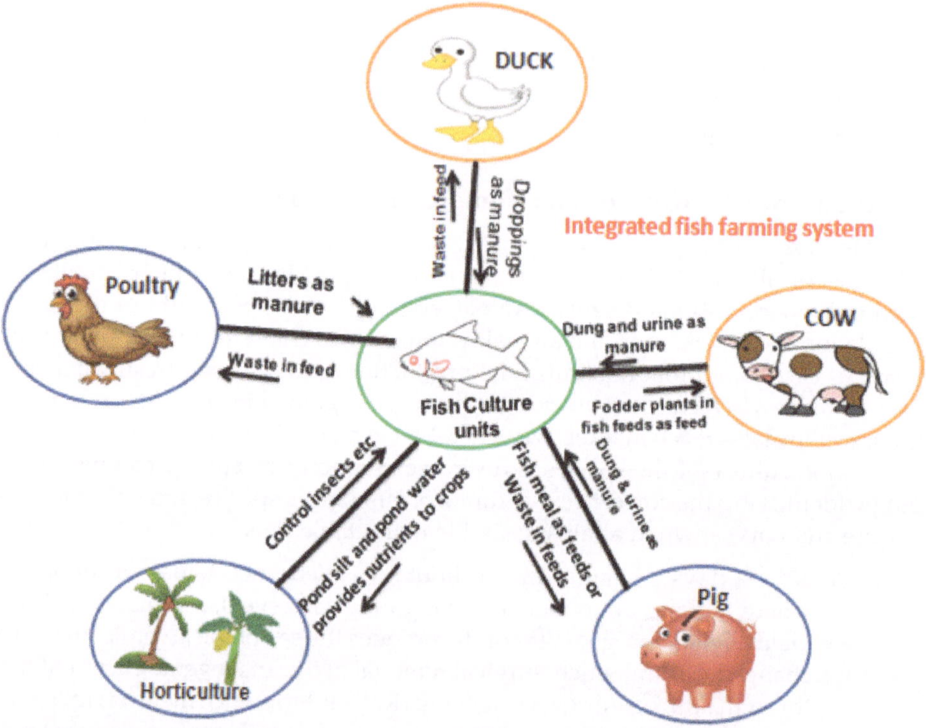

Figure 2.5. Flow Diagrammmatic Representation of Integrated Farming System.

In integrated fish farming system, the wastes or by-products obtained from one system *i.e.* minor commodity becoming an input of fish culture systems. The by products obtained from fodder plants, fruit plants and agriculture crops *i.e.* rice bran, rice polish, flour, oil cake, soybeans *etc.*, are used in varied processed forms in aquaculture as a feed input. Besides, the wastes both urine and dung from different live stocks *i.e.* cows, poultries, chickens, pigs, rabbits, goats, sheep, silk warms *etc.*, are used as a sources of manure to generate fish food organisms and helps in the sustaining of aquatic food web.

There are basically two types of IFF followed by the farmers in India. The agri-based system includes paddy-fish, mushroom-fish, seri-fish, vermi compost-fish *etc.*, with aquaculture as major component whereas other agriculture practices as minor components. The livestock fish system includes cattle-fish, pig-fish, goat or sheep-fish, duck-fish, rabbit-fish *etc.*, with an objective to increase the farm productivity maximizing synergies between these components. Thus the IFF provides a new farming approach to increase the food production in a sustainable way which ensures better biodiversity, ecosystem and integration of soil-water fertility management practices

Advantages

- ✰ Offers tremendous potential for food security and production
- ✰ Optimization the use of available natural resources
- ✰ Diversification of income generating activities
- ✰ Improvement of soil-water fertilities
- ✰ Minimize the use of chemicals (pesticides, fertilizers, antibiotics) Aquatic Biodiversity
- ✰ conservation and sustainable use could be enhanced
- ✰ Efficient utilisation of farm space for multiple productions
- ✰ Recycling of waste or by-product
- ✰ Organic in nature
- ✰ Eco-friendly
- ✰ No need to use artificial fertilizer
- ✰ Reduces the costs of production and economically

6. Integrated Multi-Trophic Aquaculture (IMTA)

IMTA is a way of farming multiple marine or freshwater organisms from different positions or nutritional levels or trophic levels in the same system at the same time (Barrington *et al.*, 2009). Integrated multi-trophic aquaculture (IMTA) is the farming of aquaculture species from different trophic levels, and with complementary ecosystem functions, in a way that allows one species' uneaten feed and wastes, nutrients, and by-products to be recaptured and converted into fertilizer, feed, and energy for the other crops, and to take advantage of synergistic interactions between species.

IMTA is the practice which combines the cultivation of fed aquaculture species (*e.g.* finfish/shrimp) with organic extractive aquaculture species (*e.g.* shellfish/herbivorous fish) and inorganic extractive aquaculture species (*e.g.* seaweed) to create balanced systems for environmental sustainability (bio mitigation) economic stability (product diversification and risk reduction) and social acceptability (better management practices). Typically, carnivorous fish or shrimp occupies the higher trophic levels of IMTA's. They excrete soluble inorganic ammonia, phosphorus (orthophosphate) and particulate organic carbons (Fish faecal matter). Seaweeds and similar species can uptake these inorganic nutrients (inorganic nitrogen and phosphorous) directly from their environment. The organic nutrients (fish faecal matters and uneaten feeds, POM) are fed by shellfishes and deposit feeders and their excretory inorganic products again uptake by the seaweeds.

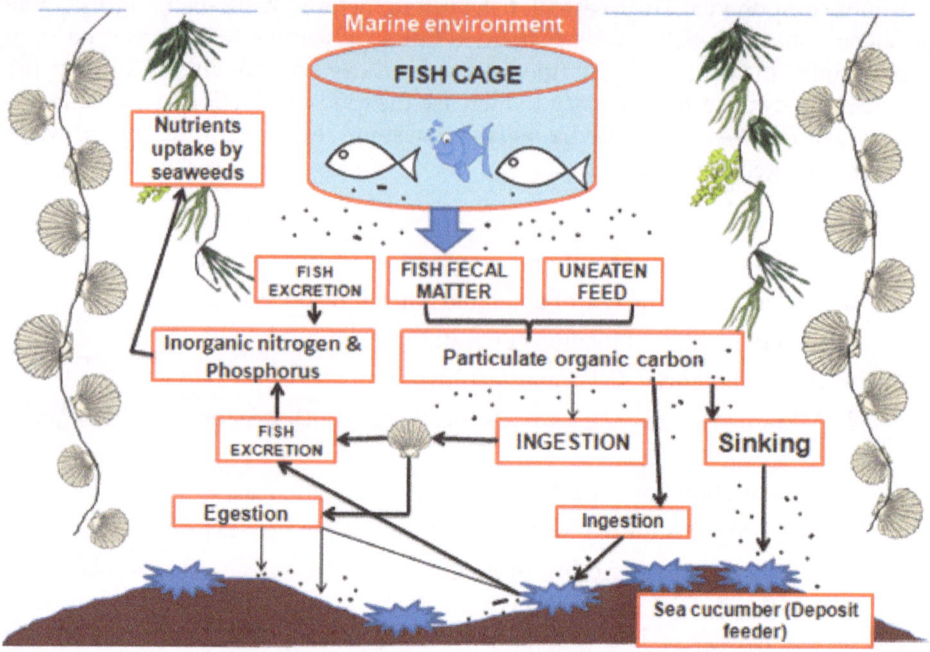

Figure 2.6. Flow Diagrammmatic Representation of Integrated Multi-Trophic Aquaculture (IMTA).

Thus the IMTA provides a new farming approach to increase the food production by efficient utilisations of trophic levels in a sustainable way which ensures better biodiversity, bio security and ecosystem. Integrated Multi-Trophic Aquaculture = Fed Aquaculture (Finfish) + Extractive Aquaculture *i.e.* Organic (Shellfish) Inorganic (Seaweed).

Advantages

- ✰ Offers tremendous potential for food production and security
- ✰ Promotes economic and environmental sustainability by converting by-products, wastes and uneaten feeds.
- ✰ Reduces the eutrophication in culture environment
- ✰ Increase of economic diversification
- ✰ Reduces negative environmental impacts
- ✰ Eco-friendly.
- ✰ Efficient utilisation of all the trophic levels

7. Organic Aquaculture

Organic aquaculture refers to the production processes and practices of ecological production management systems that promote and enhance biodiversity, biological cycles and biological activity. It is based on minimal use of off-farm inputs and on holistic management practices that restore, maintain and enhance species diversity and ecological harmony (Prein *et al.*, 2010). Moreover, the primary goal of organic aquaculture is to optimize the health and productivity of interdependent communities of soil life, plants, animals, and ensures that the production system should be socially, ecologically and economically sustainable.

Organic aquaculture is based on the principles and practices of Intensive monitoring of environmental impact, use of polyculture, Integration of natural plant communities in the farm management, use of indigenous species as far as possible, without the use of hormones, irradiation and antibiotics and appropriate stocking density for food production in a sustainable way.

Advantages

- ✰ Sustainable fish farming system
- ✰ Preserves the ecosystem
- ✰ Controlling supply and demand
- ✰ High nutritional Value
- ✰ Eco-friendly
- ✰ Creating more job opportunities
- ✰ Poison-free
- ✰ Food Tastes Better
- ✰ Lower Input Costs
- ✰ Disease Resistance

8. Compensatory Growth Technology

Compensatory growth is the phase of rapid growth, greater than normal or control growth, which occurs upon adequate refeeding following a period of under nutrition (Ali *et al.*, 2003). Compensatory growth (Stunted Fingerlings) is identified by being significantly faster than the growth rate of control fishes that have not experienced growth depression, held under comparable conditions (Biswas *et al.*, 2016). The fish to be in starvation followed by what may be termed compensatory growth once feeding is resumed. In this method, the fish that have been starved for 3 weeks and then fed for 3 weeks show a weight gain equivalent to or greater than that of fish fed normally for the 6 week period.

Advantages

- ✰ Carps are known to grow rapidly during the second year of their age.
- ✰ Higher survival rate in grow out ponds.
- ✰ Good immunity against diseases.
- ✰ More tolerant to environmental fluctuations.
- ✰ Require less time to reach marketable size (5-6 months).
- ✰ High growth rate and can be sold at higher prize.
- ✰ Higher production and productivity.

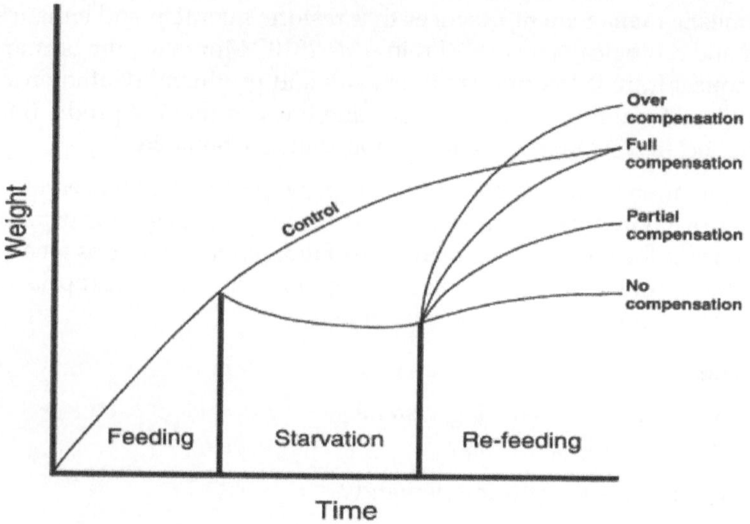

Figure 2.7. Graphical Representation of Compensatory Growth Technology.

- ☆ Unhealthy seeds are perished during stunting periods, so we get only healthy seeds.

Conclusion

These advanced aquaculture farming systems and other technological innovations are showing a positive impact on aquaculture success, productions, investment and marketing potential. The development of these farming systems in aquaculture should provide a means of producing healthy and fast growing animals, through eco-friendly means.

However, this development will largely depends on the desire, willingness, capacity building and infrastructure development of the farmers to work in collaborations with scientists and the international donor community to assist in related farming technologies.

References

1. Alok Kumar Jena, Pradyut Biswas and Himadri Saha. 2017. Advanced farming systems in aquaculture: strategies to enhance the production. Innovative Farming, 2(1): 84-89. http://nfdb.gov.in/about-indian-fisheries.htm
2. All about fish farming in India, Krishijagram.com on 06 September, 2018 4:38 PM IST By: Dr. Balwinder Singh Dhillon, Dr. A.P.S. Dhaliwal and Dr. J.S. Brar.
3. Ali, M., A. Nicieza and R.J. Wootton. 2003. Compensatory growth in fishes: a response to growth Depression. Fish and Fisheries, 4(2): 147-190.
4. Amir, S. 2013. Monosex culture of prawns through androgenic gene silencing. Infofish International, 10(4): 22-24.

5. Avnimelech, Y. 2009. Biofloc Technology — A Practical Guide Book. World Aquaculture Society pub., Baton Rouge, Louisiana, United States, pp. 182.

6. Ayyappan, S. 2011. Hand book of fisheries and Aquaculture. Indian council of Agricultural Research pub., New Delhi, India, pp. 439-448.

7. Azad, K.N. and M.A. Salam. 2016. Aquaponics in Bangladesh: current status and future prospects. Journal of Bioscience and Agriculture Research, 7(2): 669-677.

8. Barrington, K., T. Chopin and S. Robinson. 2009. Integrated multi-trophic aquaculture (IMTA) in marine temperate waters. In: FAO Fisheries and Aquaculture Technical Paper Number 529. FAO press, Rome, Italy, pp. 7–46.

9. Biswas, P., M. K. Datta, H. Saha and S. C. Mandal. 2016. Stunted Fingerlings-replacement of conventional fish seed. In: Vocational training manual on aquaculture production and its management. College of Fisheries press, Lembucherra, Central Agricultural University, Imphal, pp. 59-61.

12. Crab, R., T. Defoirdt, M. Wille, P. Bossier and W. Verstraete. 2012. Biofloc technology in aquaculture: Beneficial effects and future challenges. Aquaculture, 356: 351–356.

11. Prein, M., S. Bergleiter, M. Ballauf, D. Brister, M. Halwart, K. Hongrat, J. Kahle, T. Lasner, A. Lem, O. Lev, C. Morrison, Z. Shehadeh, A. Stamer and A.A. Wainberg. 2010. Organic aquaculture: the future of expanding niche markets. In: Proceedings of the Global Conference on Aquaculture, Phuket, Thailand, September 22-25, pp. 549–567.

Chapter 3

Breeding and Culture of Ornamental Fish and Aquarium Keeping

R. Ram Mohan

Assistant Professor (C), Department of Aquaculture, College of Fishery Science, Muthukur

Ornamental fish farming is a promising and upcoming sector that offers alternative employment to both rural and urban unemployed youth and contributes to the growth of national income and also has greater potential for earning foreign exchange. Global fish markets records were established in 1976 with just 28 countries which were known to be exporting ornamental fishes, later reaching 105 in 2004; presently more than 125 countries are involved in this trade. The global exports market was rose steadily from 2000 valued at US$ 177.7 million and reached a peak value of US$ 364.9 million in 2011. The Export value for the global ornamental fish industry in 2016 stood at US$ 337.70. Singapore has over 20 per cent share in the world market while Malaysia and Sri Lanka command 7 per cent and 5 per cent, respectively. USA is the largest market for the ornamental fishes, followed by the European Union and Japan. India's domestic trade is Rs.10 crore and is growing at the rate of 20 per cent annually. Ornamental fishes of India are contributing about 1 per cent of the total ornamental fish trade. These fishes are exported to the tune of 69.26 tons, having the value of Rupees 566.66 crores in 2016.

The ornamental species are categorized into indigenous and exotic. Availability of a vast number of native species has contributed significantly to the development of ornamental fish industry in the country. North-eastern states, West Bengal, Kerala and Tamil Nadu are blessed with potential indigenous species. About 90 per cent of native species (85 per cent are from northeast India) are collected and reared to meet export demand. Presently, nearly about 100 native species are reared as aquarium fish. There is also a great demand for exotic species due to its colour, shape and appearance. More than 300 exotic species are covered in the ornamental fish trade, but a greater demand exists for this trade. About 200 species are bred in India. 90 per cent of India's exports go from Kolkata followed by 8 per cent from Mumbai and 2 per cent from Chennai.

Aquarium Fishes

Aquarium fishes are broadly grouped into 2 categories on the basis of their breeding behaviour, *viz.*, egg-layers (oviparous) and livebearers (Ovo-viviparous). Egg-layers are further classified as egg-layers with no care (non guarders), egg-layers with care (guarders), egg buriers, mouth-incubators, nest-builders and egg-carriers on the basis of parental care.

Egg Layers

Important group of egg-layers are barbs, rasboras, goldfish, tetras, danios, betas, angelfish and gouramis. Among them barb is one of the most important groups. Barbs of Indian origin are rosy barb, striped barb and aruli barb. The major group of danios includes giant danio, pearl danio and zebra danio. Among rasboras slender rasbora, glowlight rasbora and scissortail are the important ones. Goldfishes are the most preferred fish for the aquarium keepers. Some common varieties are comet, lion head, oranda, red cap, veil tail, shubunkin, bubble eye *etc.* The fish grows up to 20 cm in length, but starts breeding when it reaches only 6 cm long.

The tetras are small fishes of 3-8 cm length that originated from South America. Most common varieties are the black widow tetra, flame tetra, neon tetra and cardinal tetra. The species, *Betta splendens*, popularly known as Siamese fighter, occurs in varied colours. The males become aggressive in presence of other males. Angelfish is an important group of aquarium fish with different varieties such as black, veil tail, marble and albino. Among gouramies, three-spot gourami, pearl gourami, moonlight gourami, giant gourami and kissing gourami are the important species.

Egg Scatterers Laying Non-adhesive Eggs

Zebra fish (*Danio spp.*) is considered as egg scatterer, which lays non-adhesive eggs. Like many aquarium fishes, zebra fish also eats away its own eggs and spawn after breeding. As the precautionary measure, the bottom of the aquarium should be loaded with a layer of round pebbles of about 6-8 cm diameter. The breeding pair has to be well fed with live food like small zooplanktons. During breeding the male female ratio should be maintained at 2:1 or 3:1. The female is introduced in the breeding tank one day earlier than the males. The eggs require 2-3 days of hatching time, if the temperature is favourable. As soon as the tiny hatchlings are seen in the aquarium tanks, the parents are to be removed. The hatchlings take 2 days to absorb their yellow yolk sac. After 2 days, they are fed with infusorians for

4 days. Subsequently rotifers and smaller zooplanktons can be fed for 1 week, after which they can be provided with powdered formulate feed.

Figure 3.1. Egg Scatters Laying Non-adhesive Eggs.

Gold fish (*Carassius* spp.) is considered as egg scatterer laying adhesive eggs. When secondary sexual characters appear, the male and female gold fishes are selected and kept in circular glass tanks (24″x 12″x 15″) or ferrocement tank (3.5 ft. x 2.5 ft.) after disinfecting the containers with 1 ppm solution of potassium permanganate. The water used should be a mixture of ground and filtered pond water. The tanks should be placed where some early morning sunshine and no sunlight afterwards fall. Since goldfish eggs are sticky in nature, they require a surface to adhere. For this various artificial nets or submerged aquatic plants such as *Hydrilla* can be used. The nets should float close to the surface of water. The water temperature should be maintained between 20°C and 30°C. The spawner and milter in the ratio of 1:2 are released into the breeding tank in the late evening hours. Egg laying usually takes place within 6 -12 hrs. The moment spawning is over, nets should be transferred to a different container, or parent fishes are removed from the breeding tank. Generally a female lays about 2000 - 3000 eggs. Healthy eggs are golden in colour and transparent in the beginning and gradually the transparency decreases. Unfertilized eggs will remain opaque. Under ideal condition, within 3 days, the eggs hatch-out with a hatching rate of 80-90 per cent. When the young larvae start to float the nets and aquarium plants can be removed.

Egg Depositors

Barbs (*Rasbora spp.*), small fishes that move in groups are ideal for a well-planted community aquarium. A temperature between 25° and 28° C is optimum for their breeding. They are difficult to breed but can lay up to 250 eggs/female. Like barbs they require soft, slightly acidic (pH 5.5) environment. After conditioning male and female are placed in a tank planted with flat leaved plants. Once spawning occurs, as indicated by the slimness of the female fish, remove both parents from the breeding tank. The eggs laid on the underside of the flat levels will hatch after 24-36 hr and the resultant hatchlings become free swimming after 3-5 days. At this stage, the tiny hatchlings should be fed with infusorians and newly hatched brine shrimp. As they grow bigger, they should be fed with zooplanktons, like *Moina* and *Daphnia*.

Egg Buriers

Among the egg buriers, the killifish (*Aplocheilus spp.*) is the most important. They lay their eggs in a soft peat at bottom of the tank or in densely planted aquarium tanks. They are good jumpers; therefore, they should be kept in covered aquarium. The eggs are capable of remaining viable even under dried condition and hatching may be possible even after some weeks or months, when placed again in water. In drought condition, parents may die but their eggs remain alive until the next rain. They rarely grow up to 3-4 cm in total length and are short lived.

Nest Builders

The common nest builders are Gourami, Siamese fighter and Angelfish. They are bubble nest builders and incubate their eggs in floating nests, specially made by the male fish.

Gourami (*Osphronemus* spp.)

Among the nest builders gouramis are the most popular. For breeding purposes males and females are kept separately in different tanks for a week and fed with live food. When abdomen of female becomes grossly, distended with eggs it is transferred to a smaller breeding tank with water level of 5-6″ at 28°-30°C. The tank should contain plenty of fine-leaf plants such as *Cabomba* and some floating plants. The water hardness of 100-200 ppm and pH of 7.0-7.5 are ideal. After 1-2 days, mature male is introduced in the breeding tank.

A transparent perforated plastic sheet or a glass is covered over the tank to keep the humidity and temperature at high level, which help to maintain the bubble nest in good condition. The male soon begins to build the bubble nest. This is possible by engulfing a large gulp of air at the water surface and converting it into many smaller bubbles that are passed into gill chamber and coated with an anti burst agent before release.

After making the nest, the female deposits a large number of eggs in the nest. After breeding, female is removed, while males guard the nest. Hatching takes place within 24-36 h and the moment fry swim freely from the nest, males are removed from the tank. The young ones are given infusorians at this stage and after a week newly hatched artemia and small cladocerans are provided. As they grow they accept all kinds of prepared feeds.

Siamese Fighter (*Betta splendens*)

Adult fish attains sexual dimorphism at a length of 6 cm. It is best to attempt breeding with fishes that are about 9-12 months old. Allow one male to every 2 or

3 females. Females should be at least the same size as the male. Males are kept in small aquaria of 2-5 litre capacity, while females are kept in tanks containing 25-50 litres of water. Another breeding tank containing 50 litres water (depth 15 cm) and having leaf plants like *Myriophyllum* and *Cabomba*, is required. The tank has to be partitioned into two using fine mesh net. In one half females and other half male fish is placed. Water temperature is maintained around 27°C. The male starts building a bubble nest quite quickly and once this is underway, the partition net is removed. At this crucial stage male should accept the female, otherwise male starts vigorous display of chasing which ultimately leads to fin tearing of female.

Figure 3.2. Betta Breeding Trap.

Fighters often spawn in early morning and within few hours 200-300 eggs are laid. As the eggs are shed and fertilized, they sink to the bottom. Males then collect them in his mouth and spit them into the bubble nest. At the end of spawning females are removed and male is left to guard the nest for 3 days after which it is removed. The eggs hatch after 36-48 h. The smaller fry become free swimming after 5 or 6 days during when they can be fed with infusorium and egg-yolk. After 3 or 4 days, frys generally accept fine dry foods. The temperature of the water should be warm at around 27° C.

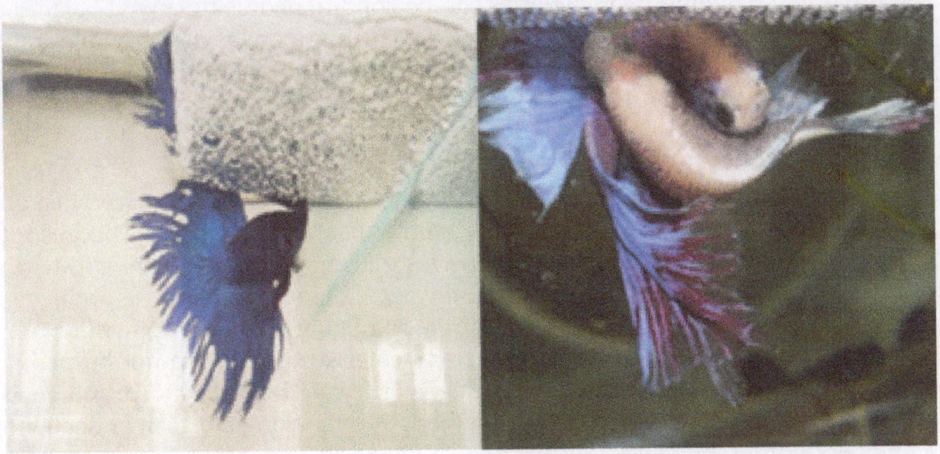

Figure 3.3. Bubble Nest and Mating Betta.

Angelfish (*Pterophyllum* spp.)

The mature angelfish having straight top and bottom fins without any bowing or bend is selected. They should be healthy, strong, robust and active. Unfortunately it is very difficult to differentiate between a male and female angelfish externally. In the beginning, 6-8 potential breeders are selected which can be set in a 100-litre tank and they are fed well with live food. The fish soon makes pairs and start displaying breeding and courtship behaviour. They spawn on broad-leaved Amazon sword plants in the aquarium. Angelfish prefers water with 6.0–8.0 pH despite 6.5-7.4 is being ideal, a water hardness of 50–130 ppm, and a temperature range of 24–30° C.

The female will deposit a line of eggs on the spawning substrate, followed by the male who will fertilize the eggs. This process will repeat itself until there are a total of 100-600 eggs. The pair will take turns maintaining a high rate of water circulation around the eggs by swimming very close to the eggs and fanning the eggs with their lateral fins. In a few days, the eggs will hatch and the fry will remain attached to the spawning substrate. During this period, the fry will not eat and will survive by consuming the remains of their yolk sacs. At one week, the fry will detach and become free-swimming. Fry can now feed on brine shrimp and after 2 weeks feed on powdered artificial feed.

Mouth-Brooders

Mouth-brooders: species that carry their eggs or larvae in their mouth.

Setting up of Breeding Tank for Egg Layers

Conditioning of Parent Fish

Before placing the parent fish together for spawning, they should be conditioned. Conditioning is feeding the fish with a variety of healthy foods to get them in top condition for spawning. Many species can be conditioned using a well-balanced flake food, though others should be conditioned on live foods such as brine shrimp, insect larvae, and flying insects. The parent fish can be separated while conditioning. This way, when the fish are reintroduced, they are eager to spawn.

Spawning Tank

Though some species readily spawn in the aquarium, the eggs or fry often do not survive because of predatory parents or other fish. Often the fry die because of unfavourable water conditions. Many species that practice brood care will harm other tank mates in attempting to guard the eggs. Because of all these problems; most aquarists who breed fish use a separate spawning tank. The spawning tank should be like hospital tank with protected heater so the fish are not burned; a slow-moving filter (sponge filter), so the eggs or fry are not sucked up; and good aeration. Depending on the spawning method, the spawning tank can be set up in a number of different ways.

Spawning Tank Set up for Egg-scatters

Because egg scatterers often eat their own eggs, the spawning tank has to be set-up so the eggs fall out of the reach of hungry parents. A 25-50 litre tank is sufficient for spawning of most egg scatterers. For egg scatterers like barbs and danios, which lay non-adhesive eggs, the spawning tank can be furnished with a substrate consisting of two layers of marbles or nylon netting just above the tank floor. As the eggs are laid, they fall through the marbles or the netting out of the reach of the parents. After spawning is over, the eggs or the parents can be removed. For egg scatterers that lay adhesive eggs like tetras, the spawning tank should be furnished with a substrate. The tank should be planted with fine-leafed plants. The eggs are laid amongst plants, and adhere to the fine-leaves. The parents should be removed after spawning.

Spawning Tank Set up for Egg-depositors

Depending on the type of egg depositor, the tank should be furnished differently. For those egg-depositors that care for their young, the parents can remain in the tank after spawning. Substrate spawners, depending on the species, should be given tanks with furnished glass panes, broad-leafed plants, or flat stones as spawning sites. Some species such as Discus and Angelfish prefer vertical surfaces. For cavity spawners, flowerpots turned on their side, coconut shells, and rocky caves are suitable spawning sites. The tank should be furnished with either live or plastic plants to give the fish a sense of security.

Spawning Tank Set up for Egg-burriers

A peat-moss substrate is one of the best substrates for egg-burying species. The peat moss can be removed after spawning and placed in a plastic bag to be stored for weeks to months (depending on the species). A new peat moss substrate can be placed in the tank for further spawning. In order to initiate hatching, the stored peat can be immersed in soft water.

Spawning Tank Set up for Mouth-brooders

Ovophile mouth-brooders can be bred in the main aquarium because the eggs are protected in the mouth cavity. However, it is better to separate mouth-brooders with eggs because of their potentially aggressive behaviour. There are no special breeding tank requirements other than the usual tank set-up for the species. Larvophile mouth-brooders should be placed in a breeding tank because the eggs are not protected in the mouth, but laid on a surface.

Spawning Tank Set up for Nest-builders

Nest-builders should be provided with material with which the nests are built. For bubble-nest builders, fine leafed and floating plants should be provided; species that build nests in the substrate should be given fine gravel or sand.

Stimulating Spawning

One of the best ways to induce fish to spawn, especially difficult-to-spawn species, is to simulate natural conditions. Among factors that encourage fish to spawn are the environment, the food, and the rainy season.

Water Conditions

The right water conditions are among the most basic requirements in spawning fish. Thus the water conditions should be similar to those in the natural environment of the species. Another important environmental condition is the right tank set-up including hiding places, spawning sites, lighting, water current and social conditions (schools).

Food

The right foods are important to encourage spawning. Without proper foods, natural conditions cannot be entirely recreated. Some of the live foods that often can make a difference in spawning success are earth worm, mosquito larvae and fruit flies.

The Rearing Tank

A rearing tank is not required with species that take care of their young, although they are still recommended. For species that do not take care of their young, the rearing tank can be the same tank as the spawning tank as long as the parents are removed. The rearing tank should have a protected heater, a sponge filter and plastic or live plants. In addition, three of the tank sides should be covered with black paper, because a light can encourage fungal infections and cause discomfort for the fry. The water in the rearing tank should be similar to the water used for spawning.

Raising the Fry

When the eggs hatch, the larvae that emerge look nothing like the parent fish. Instead, the larvae have a large, yellow yolk sac and are barely able to move, let alone swim. The larvae will feed off the egg sac until all the yolk is gone. Once the yolk sac disappears, the hungry fry will begin to look for food. The fry of small fish can be first fed infusoria, "green water," or egg yolk. Later these fry can be fed larger foods like white worms, Daphnia, Artemia, nauplii, and ground flakes. These foods are good for slightly larger fry such as those of cichlids.

Once the fish grow larger, larger foods like brine shrimp, larger Daphnia, flakes, insect larvae, and chopped Tubifex worms are accepted. 25-50 per cent of the water in the rearing tank should be changed daily. Be sure that the "new" water added has characteristics like the water taken out, because fry are more sensitive to sudden changes in the water. The fry should be fed several times a day. Many species need periodic sorting by size, so that larger fish do not cannibalize smaller fish. With favourable water conditions, regular water changes, and generous feeding, the fry should grow quickly. Unhealthy and deformed fish should be removed.

Setting up of Breeding Tank for Livebearers

Livebearers are fish that bear live young ones. There are two types of livebearers: ovoviviparous, where the eggs form and hatch within the female before birth; and viviparous, where no eggs are formed, and the young are nourished through an umbilical-like cord or from secretion by the female. Livebearers are generally prolific and are easily bred.

Breeding Livebearers

Unlike other fishes, female livebearers are fertilized internally by transferring milt into females by means of the males modified anal fin, the gonopodium. Breeding pattern of guppies and other fishes of this group vary and hence have been dealt separately.

Guppy (*Poecilia reticulate*)

Male guppy reaches up to 2.5 – 3.5 cm length, while the female is usually larger when fully grown. They thrive in a large well-planted tank with temperatures within 20°–25°C. Gravid females with their abdomen enlarged are collected and placed in the breeding tank (30 cm x 20 cm x 20 cm) individually or pairs. Plants like *Cabomba* can be placed in the aquarium. Once individual females give birth to 20-200 young ones, the tank is aerated. For mass breeding of guppy a tank size (100 cm x 100 cm x 60 cm) is ideal, where a perforated cylindrical basket can be provided in one side of the tank encircled with fibrous plastic flowing filaments where the female can drop the young ones. Soon after birth, the young ones escape from their mother and enter into the perforated basket and latter the young ones are collected from the basket and placed in separate tanks, for further rearing.

Figure 3.4. Guppy Premating Behaviour and Young Ones.

Platy, Swordtail and Molly

The platies and swordtails (*Xiphophorus spp.*) take 6-8 weeks, while mollies (*Poecilia spp.*) take 12-16 weeks to mature. Like guppies the male used to insert its gonopodium with milts into female, fish and eggs are fertilized inside the mother's body. After fertilization, the embryo grows to tiny young ones and becomes ready for free swimming within 4 weeks of gestation period. Platies, swordtails and mollies are quite hardy fish; they will breed well in most type of water, so long as it is not too soft or acidic. Addition of little aquarium marine salt or common salt to water (0.5 –1 g/litre) is beneficial for mollies.

Procedure

The preparation for the breeding tank is simple. It is need not to be too large for most species, and should be filled to a depth of about 8″ (20cm) with matured water and the temperature raised to about 25°C. The tank should be well stocked with fine leaved plants. Once the female has given birth to the full brood she can be removed.

Figure 3.5. Platy, Swordtail and Molly Fishes.

Instead of plants, nylon knitting wools can also be used. Handling livebearers when they are near the time to delivery can cause premature birth. Premature babies have not completely absorbed their yolk sac, which can be seen attached to the substrate.

Conditioning

Before placing the parent fishes together for spawning, they should first be conditioned. Conditioning is feeding the fish with a variety of healthy foods to make them attain suitable condition for spawning. Many species may be conditioned using a well-balanced flake food, though others may be conditioned with live foods such as brine shrimp, insect larvae, and earth worms. The parent fishes should be separated while conditioning and these fishes will be ready to spawn when reintroduced.

Breeding of Livebearers

Live bearing fish, the eggs are present in the egg duct where they get fertilized. Upon hatching, the fry are not immediately delivered, but they remain in the safety of the mother's body until they reach a stage of development equivalent to the young of egg layers that have absorbed the yolk sac and become free swimming. The ideal temperature must be 27°C.

As the male matures, the anal fin becomes more pointed and straightened into a rigid tube like projection, which is called gonopodium. The gonopodium is carried normally close to the body and pointing rearwards. However, it is a mobile organ and can be angled in almost any direction. The males court the females with their fins erect and they chase the female untill the opportunity presents itself for a lightening thrust of the gonopodium. The female is fertilized by the nearest touch of the gonopodium on her vent and one fertilization will last for several broods. Unlike the male, the female has normal anal fin. Females can have upto eight broods from one fertilization therefore it is unnecessary to remate after the first brood. The period of gestation is being constant for every species.

Table 3.1: Breeding Characteristics of Livebearers

Species	*Sexual Dimorphism*	*Size*	*Optimum Water Condition for Breeding*	*Gestation Period (Days)*	*Young Ones/ Females*
Guppy, *Poecilia reticulate*	Male is smaller with more flowing fins and pointed anal fin or gonopodium	Male 2.5 -3.5 cm; Female 5 -6 cm	Temp. 20-28°C, water hardness, 50-100 mg $CaCO_3$/ litre (moderately soft water)	21-35	20-100
Platy, *Xiphophorus maculates*	Male is smaller and slimmer with gonopodium. General coloration is red, gold, blue, black, brown *etc.*	Male 3-4cm; Female 4-5 cm	Temp. 23-28°C, water hardness, 50- 100 mg $CaCO_3$/ litre (moderately soft water)	28-42	10-100
Sword tail, *Xiphophorus helleri*	Male is smaller and slimmer with gonopodium and pronounced sword like projection on caudal fin	Male 6-7 cm; Female 7-9 cm	Temp. 23-28°C, water hardness, 50-100 mg $CaCO_3$/ litre (moderately soft water)	28-42	20-100
Black molly, *Poecilia sp.*	Male with gono podium, dorsal fin is flowing and bigger. Males are smaller than females and slimmer also.	Male 7-8cm; Female 9 cm	Temp. 23-28°C, Aquarium salt at 0.5-1.0 g/litre	40-70	30-70

Females about to give birth are said to be ripe. This condition can be determined by the appearance of the dark, crescent shaped area in the female body close to the vent known as the gravid spot, which is accompanied by a general fattening of the belly when viewed from above. The eggs develop and actually hatch out inside the mother and leave her body as fully developed fish. The number of young in a brood is largely dependent upon the size of the female. Irrespective of the number in the brood, all the fry are approximately in same size at birth. Livebearers are notorious cannibals; they will devour their young ones as soon as they are born. This can be prevented by having an abundance of cover for the young ones in the form of bunched fine leaved plants, or to use a breeding trap, which restricts the female to a small part of the aquarium, but allows the fry to escape into the wider reach of the breeding tank.

Preparation of breeding tank is simple. It is need not to be too large for most species, and should be filled to a depth of about 8 inch (20cm) with matured water and the temperature is raised to about 25°C. The tank should be well stocked with fine leaved plants. Once the female has given birth to the full brood she can be removed. Instead of plants, nylon knitting wools can also be used. Handling livebearers when they are near the time of delivery can cause premature birth. Premature babies have not completely absorbed their yolk sac, which can be seen attached to their bellies. Livebearers normally kept in a community aquarium tank will breed indiscriminately, and will often crossbreed between similar species. If the aquarist is interested in obtaining a particular colour strain or any other feature for that matter it is imperative that the sexes are housed separately.

After 2-3 days the female again becomes pregnant even without the contact of male. The sperm transferred during first mating is stored in the female body and when eggs are formed the sperm will join with eggs and form young ones. Thus by single mating young ones can be released 8-10 times. In a year it is possible to get young ones 10 times from one female.

Feeding of Fry

Baby fish *i.e.* fry can swim and eat from the moment they are born. It accepts infusorians, finely crushed dried food, brine shrimp naupli, *etc.*,

Ornamental Fish Culture

Ornamental fishes really are nature's wonderful creation. Ornamental fish keeping is the second most preferred hobby in the world and the number of hobbyists for ornamental fish keeping is rising day by day because it provides a great opportunity for entrepreneurship development and income generation. There are number of promotional schemes from various government agencies, which would boost up aspiring entrepreneurs. The technology is quite simple and easy to take up and attain success.

Prerequisites for Ornamental Fish Culture

The success of ornamental fish culture greatly depends on the location and layout of the unit with adopting proper management protocols in the breeding and rearing. The basic necessities are like plenty of fresh water, quality brood stock and uninterrupted electric supply. For this reason, one has to plan meticulously while establishing an ornamental unit. There are also several other parameters, which are need to be considered to have success in this culture practice. Some of these parameters are discussed below.

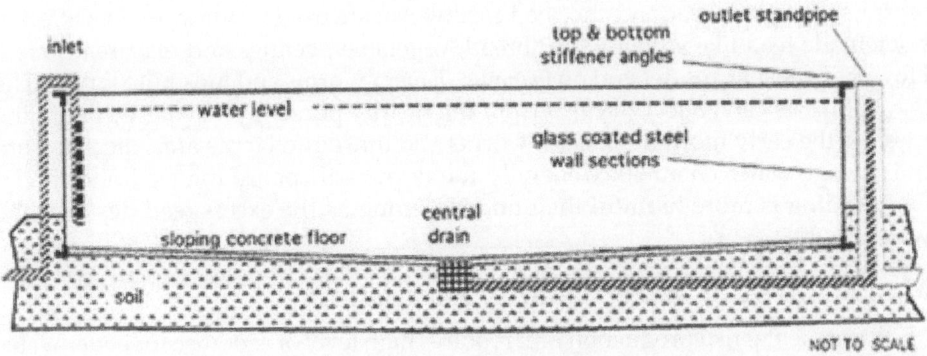

Figure 3.6. Large Scale Ornamental Fish Farm Design.

Important Ornamental Fishes

Generally, the exotic fish dominate the Indian market and more than 250 varieties have been recorded from the domestic market. More than 200 species of these freshwater fish are bred in different parts of India by the ornamental fish breeders and aquarium hobbyists. The egg layers lay adhesive or non-adhesive eggs on the glass wall or aquarium plants. Some parents show parental care and some destroy their eggs so different breeding setups are needed. In contrast, live bearer fish release young in batches and are easy to breed compared to egg-layers. Common exotic live-bearers like guppy, molly, swordtail, platy and egg layers like gold fish, koi carp, tiger barb, Siamese fighting fish, tetra, silver shark, angel, red-tailed black shark, red finned shark, *etc.* dominate the market. Native ornamental fish include honey gourami, rosy barb, zebra fish, glass fish, loach, *etc.*

- ☆ For small scale culture a minimum of 500 Sq.ft land area is required where the farmer can construct few rectangular concrete tanks. For big projects more than 1ha land area is required and a numbers of earthen ponds required in addition to cement tanks for the culture of koi carp, catfish, gourami and barbs.
- ☆ A shade facility is required above the cement tanks and anti-bird nets are required above the earthen ponds for breeding and rearing of ornamental fishes.
- ☆ Sufficient water and electricity supply at the breeding and rearing units are important.
- ☆ Selection of the quality brooder is the most important aspect for the success of the unit. Some species reproduce in soft acidic water, while a few others prefer hard alkaline water to breed.

Food and Feeding

The small-scale farmers cannot afford different readymade packed pellet feed or purchasing of artemia cyst drum, which are costly. However, they have successfully substituted low cost alternative live feeds. Water fleas, *Tubifex* or sludge worm, mosquito larvae and chopped earthworm are used by the farmers. Different homemade feed like whole-wheat bread, vegetable peelings and rice are also fed. However, most farms depend on *Daphnia*, *Tubifex* worms and mosquito larvae. The fish culturists can collect *Daphnia* from the nearby ponds by sieving through fine mesh in the early morning. *Tubifex* worms and mosquito larvae are collected from the sewage water channels. Generally the farmers dispense the feed once daily. Overfeeding is more harmful than under feeding as the excess feed destroys the water quality.

Management of Water Quality Parameters

Ornamental fish production unit requires high level of expertise for better water quality management as ornamental fishes are more sensitive to poor water quality. Many ornamental fish will perish in situations where more robust food fish species can survive. As ornamental fish are kept in tanks more numbers than their food fish counterparts, water quality is most critical. Where large numbers of fish are kept in small spaces, the build-up of nitrogenous wastes, most notably ammonia, requires the producer to implement measures to manage it properly. Regular water exchange along with proper aeration overcomes this type of problem in the tanks.

- Livebearers usually prefer to breed in hard alkaline water, while fishes like Angel, Discus, Tetra, Oscar and Loaches prefer soft acidic water. But other egg bearers like goldfish, danio, catfish, rosy barb, and fighter fish can tolerate a wide range of water.
- Bio-filter is essential for proper water quality management in the production unit.
- To increase breeding viability, avoid breeding of same fish for more than 2 years. Instead of that, fishes must be collected from far different places.
- Farm made aqua pellets can be prepared by using different agricultural by-products, for which a small pelletizer is required. The farmer can also use the commercial fish/prawn feed available in the market.

Figure 3.7. Ornamental Fish Culture in Concrete Tanks.

Preventive Health Management

Proper water quality management in ornamental fish breeding and culture is the primary preventive measures as they are very sensitive to temperature and pH. The most common diseases of ornamental fishes are reported to be white spot, mouth fungus, tail and fin rot. Some of the easily available and economic chemicals and medicines can be used as preventive measures. The easily available chemicals and medicines for health management are common salt @15-30 g/L used as bath treatment for 30 min as disinfectant, methyline blue @2.5 g/L added in aquarium water for water purification and copper sulphate or potassium permanganate @0.5-1 g/L used as bath treatment for 1 min as disinfectant.

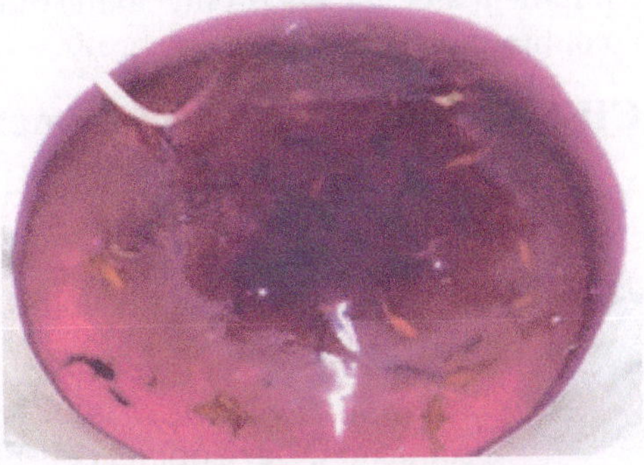

Figure 3.8. Treating with $KMnO_4$.

- ✰ Live food like Zooplankton, Tubifex, Blood worms, Chopped Earthworms *etc.* are also very much essential for brood fishes and larvae.
- ✰ For proper marketing of fish, better communication facility (roadways/rail/airport) is required nearby to the farm. Constant touch with small, middle, or large traders will benefit the farmer in marketing of his product.
- ✰ Prospective farmers can avail training and new methods demonstration to get knowledge about the new know-how in their respective field of interest and expertise. Integrated units like Ornamental fish – vermicomposite, Vegetable cultivation/backyard kitchen garden, Aquaponics can be adopted for better profit.
- ✰ Few concrete and glass tanks are to be constructed at the entry of all fish farms for proper observation and quarantine before adding to the existing stocks.
- ✰ Any unappealing situations like mass mortality or unidentified signs and symptoms among farmed fish must be attended immediately by taking advice from a specialist.
- ✰ Highly alkaline or iron containing water must be avoided as many fish don't breed in such water. For beginners in ornamental fish breeding and farming, training on the subject is a prerequisite which will help to gain knowledge and market potential. It is always better to begin with the livebearers like Guppy, Platy, Swordtail and Molly (as they breed easily) and after acquiring some experience, they can attempt for breeding and rearing of egg layers like Koi carp, gourami, goldfish, barbs, fighter fish *etc.* It is always advised for a farmer to be master on a single species than venturing into many species as they require different management practice.

Economics of a Small-Scale Breeding and Rearing Unit

The profit of ornamental breeding and rearing unit depends on the carrying capacity, candidate species, management practices and infrastructure. The marginal farmers who breed or rear the fish have to sell them earlier due to the lack of proper equipment and get less profit. On the other hand better-off farmers rear the fish to an optimum size and get more profit.

CIFA Technologies for Ornamental Fish Culture

- ✰ CIFA has successfully bred and standardized the successful rearing of 16 indigenous ornamental species from NEH, Eastern and Southern Western ghat region comprising barbs, danios, rasboras, catfish and chameleon fish.
- ✰ Recent achievement of captive breeding technology of an endangered fish *Dawkinsia tamraparneii* has been successful at CIFA leading to commercial production. Development of "Shining barb" through selection process is one of the major achievement and in ornamental fish industry, now under field trial with a private company at Udaipur, Rajasthan.
- ✰ Development of CIFACURE, a medicine for controlling fungal and bacterial diseases is gaining the momentum in many states of India. CIFA is playing major role for conducting research and training programmes on freshwater ornamental fish breeding and culture in the country by conducting several national level training programmes both in and off-campus comprising more than 2000 participants from all over the country.
- ✰ In addition to that many farmers also visit CIFA regularly for exposure to this area. Development of "ornamental fish village" through community participation is also another area of success. CIFA scientists regularly facilitates the new entrepreneurs in the development of viable business plans to establish ornamental units.
- ✰ There are several governmental incentives and subsidies to start ornamental fish farming. NFDB provides 50 per cent subsidy for establishing ornamental fish breeding and culture farms. Hence a farmer can prosper more in this direction by availing the supports laid by the Government.

Conclusion

Culture and breeding of ornamental fishes can be a promising alternative for many people as well as unemployed youths. It requires little space and less initial investment than most other forms of aquaculture. For ornamental fish farming, only a clear understanding of habits and biology of the fishes is required. It can be practiced even in urban areas with little alteration of backyard or roof of a house. As less manpower is needed, the women or the elders can run small home aquarium units and improve their social and economic upliftment.

Fabrication of Glass Aquarium

Ornamental fish tank is otherwise called an aquarium. An aquarium is a set up in which ornamental fishes of aesthetic value are displayed for recreation.

The shape of the ornamental fish tank may be circular, square, rectangular, oval, hexagonal tanks are preferred by majority of the ornamental fish hobbyist and culturists as they provide sufficient areas for free swimming of fish

Materials Required

- ☆ Glass panes (5 Nos.) with required size
- ☆ Silicone gel
- ☆ Squeezing gun
- ☆ Rope
- ☆ Thermocol

Procedure

- ☆ Aquarium tanks are generally fabricated using glass sheets. Rectangular all glass tanks are made with a glass wall thickness of 5 – 10 mm. In all glass tanks, the cut glass walls are fitted together with synthetic rubbery sealent called silicone gum.
- ☆ Prior to fabrication of the tank, sidewalls of the tank are arranged so as to have the desired shape. The glass walls are tied with a rope in order to keep the correct shape of the tank.

Figure 3.9. Fabrication of Glass Aquarium.

- ☆ Then the bottom sheet of the tank is kept flat below the arranged sidewalls so as to get the desired and correct shape of the tank.
- ☆ A cut thermocol sheet is kept at the bottom to rest the tied glass sheets temporarily. Now the silicone sealant is evenly applied all along the inner vertical and horizontal cut ends with the help of a hand applicator.
- ☆ Care must be taken to see that the joining ends of the glass pieces are free of oil, moisture or stains.
- ☆ A good sealant compound binds the glass walls in 10 to 20 minutes. However, a curing time of about 10 hrs is required to keep the set tank in position and water should be poured only after 24 hours.

Method of Construction of Glass Tanks

Flow Chart

Keep thermocol on the surface

↓

Place the bottom glass on the thermocol

↓

Apply a thick string of silicone gel along the edge of back surface

↓

Raise the back panel

↓

Apply gel in the two side edges of bottom glass pane and also edges of fixed back glass pane

↓

Place the side glasses one by one and join with other already fixed panes

↓

Smoothen the gel with finger

↓

Leave the tank for a day

↓

Check aquarium by pouring water for any leak

Seating the Tank

Soon after curing, the tank is kept on a firm wooden or slotted iron stand. The base of the stand should be even and smooth, as irregular placing of the stands would break the tank bottom. A uniform sized thermocol is also used in between the tank and stand surface to safeguard the tank from pressure.

Setting up of Aquarium

Setting up of an aquarium is otherwise called aquascaping. Aquascaping of an aquarium has two purposes.

i) To make the bare tank more attractive for the viewer.

ii) To simulate natural environment by keeping aquatic plants, rocks, gravels, *etc.*

Figure 3.10. Setting of Aquarium.

Choosing a Location

- ✰ The tank should be set up in an attractive and convenient place, which should be free from direct sunlight.
- ✰ Aquarium can be set up in home, hospitals, restaurants, hotels and other public places. The aquarium should stand on a very firm base in order to avoid jerking and toppling.

Steps Involved in Setting up of Aquarium

i) Fitting an under Gravel Filter

- ✰ The purpose of the filter is to remove dirty materials from the aquarium.
- ✰ Under gravel filters are otherwise called as biological filters.
- ✰ Toxic substances like ammonia and nitrite are converted to relatively harmless substances like nitrate by the beneficial bacteria developed on the gravel added in the biological filter.

ii) Adding the Gravels

Gravels with a particle size of 3-5 mm are used to allow good water circulation and to enable the plants to root.

iii) Installing the Air Pump

Air pump is installed to operate biological filter for airlifting and aeration.

iv) Adding Rocks

- ✰ Suitable rocks are firmly laid down on the gravel.

- ☆ The rocks are grouped for creating natural look. Several rock pieces can be stuck together with silicone gel to make a cave like structure, which is essential for maintaining cichlid fishes.

v) Filling the Tank

Tank can be filled with water through hose pipe directed over a rock in order to avoid displacement of gravels.

vi) Keeping Plants

- ☆ Tall plants are kept at the back and sides.
- ☆ The corners are filled with short species. The roots are buried in the gravel.

vii) Putting the Cover Glass into Position

- ☆ Cover glass or plastic sheets are used to cover the top.
- ☆ This type has cut-outs for cables and for feeding access.

viii) Fitting the Electrical Wiring

Electrical wiring with respect to light and filters are properly connected.

ix Setting up the Hood and Lighting

- ☆ This is the last stage before introducing the fishes.
- ☆ Fluorescent tubes are used which is fitted inside the hood.
- ☆ The hood may be made of metal, wood or fiber glass sheets.

x) Adding the Fishes

- ☆ Companionable and compatible fishes are introduced in order to avoid fighting between the fishes.
- ☆ The stocking density of fishes in ornamental fish tank is in relation to the surface area of the tank.

References

1. Ahilan, B., Felix, N. and Santhanam, R. 2008. Textbook of Aquariculture. Daya Publishing House, New Delhi.
2. Compendium on Ornamental fishes, Department of Fisheries, Govt. of Andhra Pradesh. http://www.factfish.com/statistic/ornamental per cent 20fish per cent 2C per cent 20l
3. Culture of Ornamental Fishes and Entrepreneurship Development, Sagar C. Mandal,College of Fisheries, Central Agricultural University, Lembucherra, Tripura-799210, India
4. Dey VK. The global trade in ornamental fish, Info fish International, 2016, 4/2016, www.infofish.org

5. http://mpeda.gov.in/MPEDA/cms.php?id=b3JuYW1lbnR hbC1maXNoLWV 4cG9ydGVycy1jb250YWN0LWRldG FpbHM=#.
6. ICAR - Hand book on Ornamental fishes
7. Major Exporters in India, MPEDA 2018
8. Mogalekar HS, Jawahar P. Freshwater ornamental fish diversity of Tamil Nadu, J Inland Fish. Soc. India. 2015; 47(2):27-37.
9. Ornamental Fish Breeding, Culture And Trade by B. K. Mahapatra, S. Datta, G. H. Pailan, P. Sardar A. R. Sen and S. Dasgupta
10. Ornamental fish culture leaflet, ICAR-CIFA.
11. Ornamental fish production and management, online e-learning on Agricultural Education
12. Raja S, Babu TD, Nammalwar P, Thomson Jacob C, Dinesh KPB. Potential of ornamental fish culture and marketing strategies for future prospects in India, International Journal of Biosciences and Nano science 2014; 1(5):119-125.
13. Swain SK., Sarangi, N. and Ayyappan S. 2010. Ornamental fish farming. Indian Council of Agricultural Research, New Delhi.

Chapter 4

All Male Fingerling Production of *Red Tilapia*

P. Chidambaram

Associate Professor and Head, Department of Aquaculture,
Dr. M.G.R. Fisheries College and Research Institute,
Tamil Nadu Dr. J. Jayalalithaa Fisheries University, , Ponneri

There are about 70 species of tilapia available as commercially important species for aquaculture practices. However, only few species are dominating the market *viz.*, Nile tilapia, GIFT tilapia and Red tilapia. Quality seed availability from quality brooders is the first and foremost criteria for commercial aquaculture activities. Hence production of all male tilapia fingerlings in hatcheries is necessary.

Brooders

Brooders are matured fish which will be used for breeding and seed production purposes. Tilapia is a sexually dimorphic fish where male and female fish exhibit different secondary sexual characteristics such as body weight, size, colour, *etc.* It is important to select good quality sexually matured male and female fish for captive breeding to obtain good quality seed. The following are the characteristics to identify male and female red tilapia;

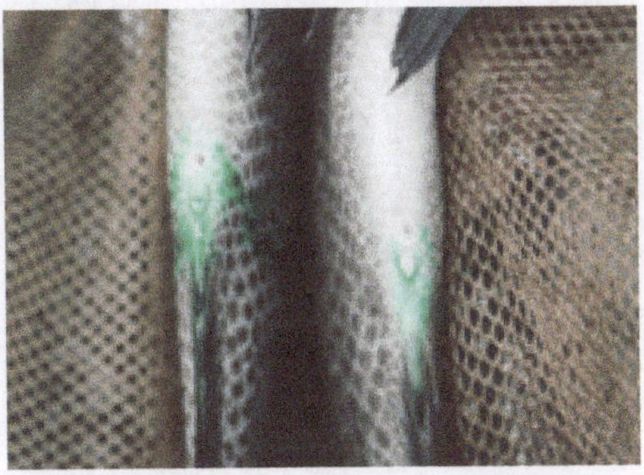

Figure 4.1. Tilapia Male and Female.

Table 4.1. Identification of Male and Female (Red Tilapia)

Male	*Female*
Gonopodium is pointed	Gonopodium is rounded
Genital papilla has only one opening (both milt and urine released in same opening)	Genital papilla has two openings (one is for urinary excretion, the other for egg release).
Pale in colour	Bright in colour
Grow bigger than the female	Normally smaller in size
Scales are rough	Scales are Smooth
Display a nest building behavior	Do not build nest.

Hapa in Pond Spawning System

Figure 4.2. Hapa Installation.

Tilapia is a synchronous breeder. Spawning occurs throughout the year in the tropics and during the warm season in the subtropics. Breeding is conducted in ponds, tanks or hapas. The common stocking ratio of female to male is 3:1. The brood fish stocking rate is variable, ranging from 0.3-0.7 kg/m^2 in small tanks and 0.2 - 0.3 kg/m^2 in ponds. The popular hapa-in-pond spawning system uses 100 g brood fish stocked at 0.7 kg/m^2. Size of spawning ponds are generally 2000 m^2 or smaller. The common hapa size is 10x3x1 m which can hold 85 brooders of 100 to 150 g each.

Brood Fish in Pond and Fry Collection

Brood fish are given high quality feed (40 per cent protein) at 1-2 per cent of body weight daily. Swim-up fry gather at the edge of a tank or pond and can be collected with fine-mesh nets. Fry collection can begin 10 to 15 days after stocking. Multiple harvests (six times per day at 5 day intervals) are conducted up to a maximum of 8-10 weeks before pond drainage and a complete harvest is necessary. Tanks must be drained and recycled every 1-2 months because escaped fry are

Figure 4.3. Broods in Hapa.

Figure 4.4: Egg Collection.

very predaceous on fry from subsequent spawns. Alternatively tanks or ponds are harvested completely after a 2-4 week spawning period. Production rate of optimum-sized (<14 mm) fry ranges from 1.5 to 2.5 fry/m^2/day.

Figure 4.5: Juvenlines of Red Tilapia.

Egg Collection and Incubation in Jar Method

- ☆ Brooders are stocked in the Hapa in male and female ratio of 1:2.
- ☆ Tilapia is a mouth brooder. In which naturally female fish will carry their eggs in her mouth.
- ☆ The eggs are collected by gentle opening of mouth and operculum then dipping the head into the water filled container.
- ☆ Eggs will be collected from the female fish mouth. The collected eggs will be incubated in the jar hatchery system.

Incubation of Eggs in Jars

- ☆ Jar hatchery system is the one of the best type of hatchery for incubating tilapia eggs.
- ☆ It is mostly made up of glass and size is 150 x 5 mm in diameter.
- ☆ It is placed at above 1m height from ground level.
- ☆ Water holding capacity of this jar is 5.5 L.
- ☆ Water flow is important because it maintains the floatation of eggs. Water flow is to be maintained at the rate of 3000 to 3200 ml per minute from bottom to top of jar.
- ☆ In each jar 250 ml of eggs can be incubated.
- ☆ After 2 to 3 days eggs of incubation, eggs will hatch out. If egg colour is brown it will hatch in one day. If it is yellow colour, it takes 2 to 3 days.
- ☆ After hatching fry will swim along with water through the pipe fitted in the centre of the jar and collected in the trays.
- ☆ Fry will be kept in the trays until yolk sac absorption is completed then fries are stocked in the fry rearing tanks.

All Male Tilapia through Sex Reversal

Commercial tilapia production generally requires the use of male mono sex populations. Male tilapia grows approximately twice as fast as females. Therefore, mixed-sex populations develop disparity among harvested fish, which affects marketability. Moreover, the presence of female tilapia leads to uncontrolled reproduction, excessive recruitment of fingerlings, competition for food, and stunting of the original stock, which may not reach marketable size. Tilapia can sexually be differentiated for several days after yolk sac absorption. If female tilapia receives a male sex hormone (17 α-methyl testosterone) in their feed, they will develop as phenotypic males. Fry collected from breeding facilities need to be graded through 3.2 mm mesh material to remove fish that are >14 mm, which are too old for successful sex reversal. The fry collected from jar are called Swim-up fry are generally <9 mm. The swim up fry will be given hormone incorporated feed for 28 days at 60mg/kg of feed.

Preparation of Hormone Incorporated Feed

Methyl Testosterone (MT) is added to a powdered commercial feed or powdered fish meal, containing >40 percent protein. Hormone incorporated feed is prepared by dissolving hormone in 95 -100 per cent ethanol, which is mixed with the feed to create a concentration of 60 mg MT/kg feed after the alcohol has evaporated. The alcohol carrier is usually added at 200 ml/kg feed and mixed thoroughly until all the feed is moist. The moist feed is air dried out of direct sunlight, or stirred in a mixer until dried, and then stored under dark, dry conditions. Androgens break down when exposed to sunlight or high temperatures. Thus the feed should be stored in cool dry and dark place to prevent inactivation of hormone.

Feeding the Fry

Swim up Fry are stocked at the rate of 3,000 to 4,000/m^2 in hapas or tanks with water exchange. If good water quality is maintained, stocking density can be increased upto 20,000/m^2. An initial feeding rate of 20-30 per cent body weight per day is gradually decreased to 10-20 per cent by the end of a 3 to 4 week of sex-reversal period. Rations are adjusted daily, and feed is administered four or more times per day. If sex-reversal is conducted in hapas, the feed must be of a consistency that allows it to float. Otherwise considerable amount of feed would be lost as it settles through the bottom of the hapa. Sex-reversed fry reach an average of 0.2 g after 3 weeks and 0.4 g after 4 weeks. The average efficacy of sex-reversal ranges from 95 to 100 per cent depending on the intensity of management.

Fry Rearing Ponds

On completion of hormone administration through feed in the indoor nursery the fry are then transferred to closed hapas in ponds. The fry are retained in these hapas until the desired market size of 2 cm to 5 cm is achieved. Hapas of 3m x 6m x 1m or 3m x 10m x 1m are suitable for rearing fry. Sufficient spacing is to be provided between hapas to facilitate free water circulation between hapas, easy feeding and easy harvest of fry for marketing

Rearing of Sex Reversed Fingerlings

Rearing of fingerlings can be carried out in open ponds in cages and in recireculatory system. After sex-reversal, fry are generally nursed to an advanced fry size before they are stocked into grow-out facilities. This procedure increases survival in the grow-out stage and utilizes growing space more efficiently. Sex-reversed fingerlings are stocked at approximately 20-25 fish/m^2 in small ponds and cultured for 2-3 months to an average size of 30-40 g. In ponds, fingerlings are given extruded feed (30 per cent protein) at a rate of 8-15 per cent of biomass per day initially, which is gradually decreased to a final rate of 4-9 per cent per day. The advanced fingerlings or smaller tilapia can be supplied to the farmers for stocking in grow-out system. In cage rearing method, series of small cages (<4 m^3) with increasing mesh size can be used to rear up to advanced fingerlings. Sex-reversed fingerlings can be stocked at a rate of 3 000 fish/m^3 and grown for 6 weeks until they attain average size of 10 g. Fish of this size can be restocked at 2 500 fish/m^3 to produce 25-30 g fingerlings in 4 weeks. These fish can be stocked at 1 500 fish/m^3 to produce 50-60 g fingerlings in 4 weeks. After four weeks the advanced fingerlings are supplied to the farmers for stocking in grow-out system. Fingerlings can also be reared in RAS system. A recirculation system stocked at 1000 fish/m^3 will produce 50 g fingerlings in 12 weeks. Fingerlings should be fed 3-4 times daily.

Packing and Transport of Tilapia Fingerlings

When the advanced fry reaches 2 to 5 cm, they can be sold to the farmers for commercial production of mono sex tilapia. The entrepreneurs has to get license from the concerned regulating authority for purchase of fingerlings from the hatchery.

References

1. Ashoke Kumar Sarker, Gopal Chandra Datta, Md. Abdur Razzak, Md. Hasnal Alam, Biswajit Mondal and Md. Sarwardy (2011). Training Manual on Improved Tilapia Culture and Dyke Cropping in Pond/Gher. A Course Manual for Tilapia Farmers Published by Director World Fish Center.
2. Breeding Seed Production and Grow-out farming of All Male GIFT in India" -Training manual Published by RGCA. A society under Marine products Export Development Authority, Ministry of commerce and industry, Govt. of India
3. GIFT Technology Manual: An aid to Tilapia selective breeding, (2004). edited by N. Puttaraksar, Published by World Fish Center, PO Box 500 GPO10670 Penang, Malaysia.
4. M.G. Hussain (2004). Farming of Tilapia - Breeding Plans, Mass Seed Production and Aquaculture Techniques. First Edition, Published by Habiba Akter Hussain -55 Kristawpur, Mymensingh 2200 Bangladesh.
5. Noor Khan, Muhammad Ashraf, Muhammad Sharif Mughal, Naureen Aziz Qureshi, Muhammadnaeem Khan, Fayyaz Rasool, Muhammad Hafeez-Ur-Rehman, Muhammad Nasir, Waqas Ali And khalid Javed Iqbal (2014). Survival and Growth Potential of Genetically Male Tilapia (GMT) Fry in Flow Through System Under Different Dietary Protein Concentrations Pakistan *J. Zool.*, vol. 46 (2), pp. 377-382, 2014.
6. Yang Yi,C. Kwei Lin and James S. Diana (2008). A manual of fertilization and supplemental feeding strategies for small-scale Nile tilapia culture in ponds, Aquaculture and Aquatic Resources Management Aquaculture CRSP Management Office, College of Agricultural Sciences, Oregon State University, 418 Snell Hall, Corvallis, Oregon 97331-1643

Chapter 5

Culture of Mud Crab Scylla Species

Ravindra Sontakke[1] and Harsha Haridas[2]

[1]Assistant Professor (C), College of Fishery Science,
S.V. Veterinary University, Muthkur, Nellore District, A.P.
[2]Scientist, ICAR-Central Island Agricultural Research Institute,
Port Blair, Andaman and Nicobar Islands

Introduction

Mud crab is an economically as well as commercially valuable portunid crab species of India prevalents in different waterbodies like coastal lagoons, estuaries and nearshore waters. Recently an attempt was made by the RGCA genetics team, to make out the taxonomic uncertainty of commonly available mud crab species in Indian coastal waters. Multiple molecular genetic markers had been used to determine the availability of two species of mud crabs, *Scylla Serrata* and *Scylla olivaccea* in India (Thampisam raj *et al.*, 2015).

India has a coastline of about 7516 km with coastal backwaters, brackishwater lakes, estuaries and intertidal swampy areas covering both the east and west coasts. It is considered that the southern part of the coast having more potential than the northern part. Out of the total 1,190,900 ha brackish water areas in India, the established area for shrimp farming is 167,193 ha which will be suitable for crab farming as well. As of states concerned, Andhra Pradesh recorded more than 2,833 ha of mud crab culture area covering nearly 50 villages. Apart from shrimp culture, farming, fattening and soft-shell production of mud crab have emerged as feasible business ventures in India (Thampisam raj *et al.*, 2015).

From past years, a decline of mud crab populations in the natural habitat throughout Indian coastal waters due to overexploitation and indiscriminate fishing of juvenile crabs by artisanal fishers, was noticed. To meet the demand of culture operations, wild seeds are collected throughout the year in Odisha (Chilka Lake); West Bengal (backwater zones of Sundarbans, Kakdwip and Namkhana); Andhra Pradesh (coastal waters of Kakinada, Visakhapatnam and Rajahmundry); Tamil Nadu (Pulicat Lake, Killai backwaters, Muthupet saline swamps, Punnakayal estuarine complex and Colachel coastal waters); Kerala (Neendakara, Cochin and Kozhikode backwaters) (Thampisam raj *et al*., 2015).

The production of good quality hatchery seed is required to meet the demand of farmers. Very few mud crab hatcheries are functioning in the world especially in the countries like Philippines, Vietnam, China. In order to increase the production of Mud Crab and facilitate a sustainable Mud Crab source for export, the research and development work. A hatchery for Mud Crab seeds production has been established by Rajiv Gandhi Centre for Aquaculture (RGCA), Ministry of Commerce and Industry, at Thoduvai Village, Sirkali, Nagapattinam, Tamil Nadu.

This hatchery aims to produce mud crab seed regularly and their rearing at its demonstration farm upto crablet sizes and also supply them to farmers as well as different research institutes like Central Institute of Brackishwater Aquaculture (CIBA) and the National Institute of Ocean Technology (NIOT). Another important activity carried out by this hatchery is the ranching of crablets in estuarine and mangrove areas of Tamil Nadu for natural stock enhancement.

Mud Crabs as Valuable Species

Mud crabs stand on a good position in both capture and culture fisheries due to their great demand, larger size and higher price. It is considered as an important component in the traditional fish or shrimp farming system of Indian and Southeast Asian countries. These mud crabs are also found in a considerable amount in traditional paddycum-fish or shrimp farming during the final harvest. Likewise, the monoculture practices of mud crabs farming are exercised in many Southeast Asian countries due to their high demand. Mud crabs can be alive after capture for a period of 5 to 8 days because it posses larger gills which makes them capable to hold a massive amount of moisture that helps to continue respiration and remain alive.

Species of Culture Importance

The identification of species and their nomenclature in mud crabs are considered as controversial subject. In the genus Scylla, four species, namely *Scylla serrata, S. tranquebarica, S. olivaccea* and *S. paramamosain are* recognized by Australians and Philippinos. It is concluded that there are only two mud crab species of genus *Scylla* found in Indian waters.

Among these two species, one species is *Scylla tranquebarica* which can grow to a maximum size of 2.3 kg. Another species is *Scylla serrata,*which grows to a maximum weight of 0.7 kg. These two Indian species can be distinguished with certain morphological characters. The polygonal markings on chelipeds, walking and swimming legs and two spines on the outer margin of the wrist of chelipeds are the characteristics features to identify the larger species, *Scylla tranquebarica*. On the other hand, the presence of only one blunt spine on the outer margin of the wrist of chelipeds and no polygonal marking on any of its legs is the distinguishing characters of smaller species, *Scylla serrata*.

Scylla serrata ***Scylla tranquebarica***

Scylla paramamosain ***Scylla olivacea***

Figure 5.1. Mud Crab Species (Courtesy: Keenan *et al.*, 1999).

Though, both the species having different life habits, they dwell in the same habitat. The species, *Scylla tranquebarica* keeps themselves buried under sand or mud while the *Scylla serrata* create deep holes at the bottom of an estuary or in earthen bunds of canals/ponds. These differences in habits make them available in larger quantities in specified habitats. For example, the larger species are most abundant in the open inshore seas/backwaters and coastal lakes, while the smaller species are abundant in mangrove areas.

Life Habits

Mud crab can tolerate a wide range of water salinities from 5 to 33 ppt and considered as euryhaline species. Optimum salinity varies from 13 to 26 ppt. They start digging holes to survive in adverse environmental conditions especially when salinity drops below 7 ppt. The required optimum temperature ranges from 18°C to 32°C. When the water temperature drops below 18°C, their feeding rate decreases and if the temperature drops as low as 12°C, crabs are accoustomed to survive in holes. They stop feeding and become dormant as the water temperature continues to drop to 7°C (www.fao.org).

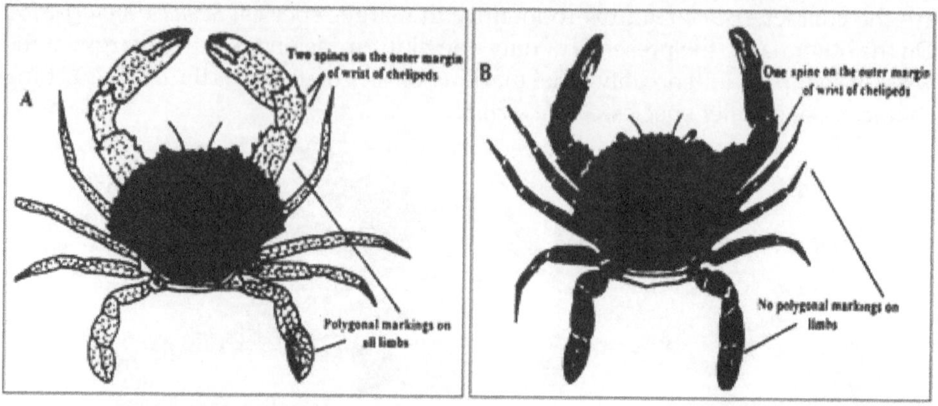

Scylla tranquebarica *Scylla serrata*

Source: Training manual on mud crab breeding and culture, CIBA.

Figure 5.2. Mud Crab Species.

These crabs moults 13 times in their life span and these moulting occurs in different stages *viz.*, 6 times each during larval stage and grow-out stage (six times) and rest one is during reproduction. Moulting occurs only at the water temperature of 15°C and preferably when it is above 18°C. While moulting, their oxygen consumption is higher r due to rapid breathing. Newly moulted crabs can not swim and lay on the bottom for two to three hours. *Scylla serrata* is a carnivorous animal preferring small mollusks, trash fish and other crustaceans (Iansan Jia and Jiaxin Chen, 2001).

Sexual Identity

The sexes of crabs can be distinguished from external characters. A slender and triangular-shaped abdominal flap on the ventral side of the body can be found in both immature and mature males of mud crabs. While a broad and triangular-shaped abdominal flap in the immature females and a semi-circular shaped flap in the matured females are present. Besides, the male consists of larger chelae and a trimmed body contour as compared to females. The two pleopods on the first and second abdominal segments are modified as copulatory organs in males. The first four abdominal segments in females comprise a biramous pleopod which carries setae for attachments of eggs for brooding.

Age at First Maturity

In natural conditions, mud crabs attain sexual maturity between 18 and 24 months. Usually, they reach the reproductive stage at a shell width larger than 7.8 cm and over 100 g of body weight. Typically, females with 8.5 cm shell width and 130 g body weight is a little bigger than males.

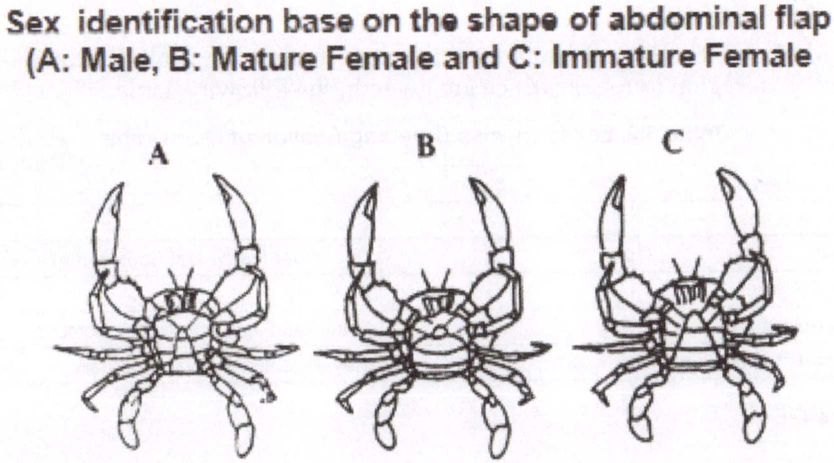

Figure 5.3. Sexual Dimorphism of Mud Crab.
Source: CIBA

The size at first maturity for females of two species of mud crabs recorded by several authors is given below.

Table 5.1. Location-wise Maturity Sizes of Mud Crabs

Locality	*Size at first maturity (CW in mm)*	
	S. tranquebarica	*S. serrata*
Karwar	—	80
SW coast	129	—
Cochin backwater	120	85
Cochin backwater	140	98
Lake Pulicat	123	83
Tuticorin	127	—

Source: CIBA.

Fecundity

The number of eggs varies with species and size of the mud crabs and the eggs are found attached to the pleopods. At the southwest coast of India, the female mud crabs with 0.3 to 0.5 million eggs attached in the pleopods are reported, while the mud crabs of Cochin backwater have 0.6 to 1.4 million eggs attached to the pleopods. *S. tranquebarica* amounting to approximately 2.5 to 7.0 million eggs and 0.05 to 2.0 million eggs in Cochin backwater and Karwar. Experiments on captive broodstock and that on the induction of maturity indicated that fecundity is 2.0 to 5.0 million eggs in the case of *S. tranquebarica* and 1.0 to 3.0 million for *S. serrata*.

Breeding Season

Mud crab can breed throughout the year though the peak breeding season varies from region to region which are given in the following table.

Table 5.2. Location-wise Breeding Season of Mud Crabs

Locality	*Period*	*Peak season*
Southwest coast of India	Throughout the year	September-February
Karwar		December-March and September-November
Tuticorin coast		April-July
Lake Pulicat		March-April and September-October
Kakinada region		May-June and October-February

Source: CIBA.

Seed Collection

Crab seeds of all sizes are available in wild. Estuaries, lakes, backwaters, creeks, mangroves and saltwater lagoons are a good source of collection for juvenile crabs by using bamboo traps, lift nets or scissors nets. A hatchery facility is available at the Central Marine Fisheries Research Institute, Kochi for commercial production of crab seed.

Transportation and Stocking

Farmers of mud crab are dependent on the wild source of small crabs or juveniles (25-50 g) collected from inter tidal flats, estuaries and mangrove to stock grow-out ponds. Crab juveniles can be transported easily by using cane basket, carton lined with moist seaweeds or mangrove leaves. Chelae are tied to prevent fighting among crabs. Mud crabs can live 2-18 days without water or in the air when packed with moist marine algae, cotton or wood shaving. Seeds having intact appendages, without injury and uniform size should be used for stocking. A difference in size leads to cannibalism. Seeds should be stocked early morning or late evening preferably night when the water temperature is considerably low. Stocking density in mud crab culture is generally far less than the shrimp farming.

Mud crab farming is an excellent option for the livelihood of coastal fisher-folk, working as associations and self-help groups in the coastal areas especially after the tsunami devastation. Mud crab farming includes nursery rearing, soft-shell crab rearing, grow-out culture and fattening. Nursery rearing involves rearing of Megalopa (0.003 g)/crablets (0.03 g) to juveniles in two phases, namely, up to 3 g (inhapas) and 3 g to 25 g (in nursery ponds). Soft shell crab rearing includes the stocking of juvenile crabs (50 to 100 g of *S. serrata* and *S. tranquebarica*) individually in a small plastic cage till they moult. Grow-out culture refers to the farming of nursery reared juveniles of about 25 g (seed crabs of *S. tranquebarica* and *S. serrata*) for 3-6 months to produce marketable size crabs. Several moultings (shedding of an old skeleton) can be observed during the growth period of crabs. Fattening refers to the holding of marketable sized crabs for about 3 to 4 weeks,during which period; the stocked crabs gain weight in the case of newly molted crabs of *S.tranquebarica* and attain full development of ovary in the case of adult females of *S serrata*. At present, all these practices are carried out in the brackishwater ponds. Fattening can also be practiced in pens/net cages in open backwaters.

Site Selection for Mud Crab Farming

Site selection is a deciding factor for the success or failure of the mud crab farming. The mandatory guidelines required for the site selection of mud crab pond construction is similar to that of shrimp culture. Already constructed aquaculture farms or abandoned shrimp farms are found suitable for crab farming. Salt-pans reservoirs at the initial stage can also be used for mud crab farming without affecting their environment. The existing regulatory guidelines include no new farms to be established within 200m from high tide line, non-use of agricultural lands and mangroves and provision of various buffer-zones between different land types such as 100 m from mangroves and drinking water resources.

In addition to the guidelines, other parameters should be addressed. The first parameter of prime concern is physic-chemical conditions that command whether a species can thrive in an environment (Temperature, salinity, oxygen, currents, pollution, algal blooms, and exchange). From a safety outlook, the water quality, flow rate and temperature should be appropriate for the species concerned. The other factors that must be considered for successful site selection are weather, shelter, depth and substrate, while the third is related to establishment of a farm and profitability (legal aspects, access, land-based facilities, security,economic and social considerations).

Environmental Conditions for Site Selection for Mud Crabs

When selecting a site, the optimum temperature and salinity levels are considered as prime criteria in mud crab culture. If any variations in these parameters can adversely affect the feeding, food conversion and growth of the mud crabs. Generally, mud crabs can tolerate a wide range of temperatures and salinity. They can withstand water temperature of 12-35°C and salinity range of 2-43 ppt, but their activity and feeding decrease rapidly when the temperature is below 20°C. Crabs can utilize the oxygen from the air. They leave the water with depleted oxygen and die at the severe conditions of hypoxia. Other parameters include available sources of stocking materials, adequate supply of inexpensive trash fish, available storage facilities for fresh, raw or unprocessed feed and security from poachers. Good quality water directly from the sea or river should be adequate all year round. The optimum range of water quality parameters is given below.

Table 5.3. Optimum Water Condition Required of Mud Crab Culture

Category	*Range*
Water temperature	23 – 32°C
Water salinity	15 - 30 ppt
DO concentration	>4
pH	8.0-8.5
Water depth (cm)	> 80

The seawater should be sufficient to fill the depth of water to atleast 60 cm height during the high tide. To determine the pond bottom elevation including the dike and gate construction, the engineering parameters should be used. The freshwater source is very much required to prevent the increase in pond water salinity during long summer months. It will help the farmer to adjust the salinity at a level favorable for the growth of mud crabs. Another socio-economic factor such as skilled labour, market accessibility as well as law and order situation in the local region should be considered.

Soil

The soil types such as silty clay, clay loam, or clayey soil with enough layers of clean mud are found appropriate for mud crab pond. This kind of soil is favorable and can meet the biological requirements of mud crabs during moulting or post-moult soft stage or feeding. This type of soil can also retain desirable water depth for mud crabs.

Criteria for Pen Culture in Mangrove Areas

Mangrove areas with a sufficient supply of marine or brackishwater throughout the year are suitable. Water depth at high tide, salinity and temperature should range between 0.3 to 1.0 m, 10 to 35 ppt and 25 to 30°C respectively. The site must be free from waves and pollution, protected from environmental hazards such as big waves and floods and secure from trespassers and poachers.

Design of Monoculture Crab Ponds

A typical brackishwater pond designed for milkfish or shrimp culture can be used for mud crab culture as well. Either new or partially developed fish ponds with water control structures can be used by maintaining the required water depth of 80-100 cm. Based on the site conditions, tidal fed or pump fed can be used. Sandy, clay or loam soil with organic matter base and preferably alkaline is recommended. Good quality water coming directly from the sea or tidal river should be adequate all year round.

Pond Shape, Size and Depth

Rectangular earthen ponds shape with areas ranging from 250 m^2 to 0.4 ha could be used. The broader axis of the pond should face the backwater canal in order to have a greater tidal effect through either wooden or brick-mortar made sluice. The water spread area of the pond should be 0.1 to 0.4 ha.

The soil removed from the bottom of the pond should be used to make the earthen dikes. If the soil is sandy, clay soil from elsewhere should be brought in for the construction of the dikes which will help in retaining the water during the culture period as well as reducing the seepage of pond water. The optimum water depth should be 80 to 100 cm.

Water Supply and Drainage

Water exchange is usually done by using the tidal effect. The in-flow and out-flow of the tidal water can be regulated by using the sluice gates. The sluice gates are fixed with bamboo screens to prevent the escape of crabs. In regions where tidal influence is less, seawater can be pumped in. The double gate system made up of wood or concrete should be preferred in the pond.

Supply and drainage must be separate to facilitate good water exchange. A leveled and cleaned pond bottom is necessary to carry out easy harvest. In the case where there is no tidal effect, a diesel or electric pump should be installed for drawl and draining of water.

Hideouts

A total of about 12 earthen mounds ($5m^3$) should be installed in strategic areas of the one-hectare pond. These mounds can also be called breathing spots as mud crab could climb over them especially during the times of low oxygen pressure, as well as the places where they can burrow. These mounds must be installed in the middle of the pond at such height so that its peaks remain above water even when the depth of 80 to 100 cm is reached. Sawed-off bamboos or already used PVC pipes (50 cm long with 15 cm diameter opening at both ends) can be added to reduce the mortality rate caused by fighting and cannibalism among equivalent crabs. These can also be provided as shelters for molting and post-moult crab (Baliao *et al.*, 1999).

Fencing the Pond

In order to prevent the escape of crabs, the area is fenced either by bamboo or nylon net (1-2 cm mesh size) which should extend about 30 cm above the waterline. The vertical bamboo or wooden posts and some horizontal bamboo splits embedded about 50 to 70 cm along the base of the dike is used to support the fence. A 30 cm wide plastic strip/sheet should be installed along the top edge of the net fence to prevent mud crab from climbing over the top. For concrete-lined ponds with relatively steeper slopes, a nylon net fence is no longer necessary. Fencing can be done on the inner periphery of the pond or the top of the dike. Catwalks may be provided for feed monitoring and stock sampling.

Design of Pond for Nursery Rearing

The hatchery-reared larvae are reared in nurseries in ponds and cages. For the larger species, there are two phases in nurserymate, one related to the rearing of megalopa to 3-gram size (nursery phase I) and the other 3-gram sized baby crabs to 25-gram sized juveniles (nursery phase II). The megalopa larvae are reared either in hapa (size 6 square meters) stationed in an open backwater or nursery ponds (80 to 100 square meters) for one month on an animal diet. Appropriate shelters (seaweeds) are provided in hapa/pond to reduce cannibalism. The average size of a single megalopa is 0.003 gram which increases to 3 grams in one-month rearing. The three-gram sized baby crabs are stocked in nursery ponds and reared on a diet of trash fish for 45 days, by which time, they attain 25 grams. The expected survival is 60 to 80 per cent in nursery phase I and 50 to 60 per cent in nursery phase II. The nursery ponds should be 100 to 200 sq.meters. The design and construction are similar to the grow-out ponds.

Design of Pond for Crab Fattening

Generally, crab fattening can be done in ponds, cages or pens. For this purpose, the pond should be small in size ranging from 0.1 to 0.5 ha with a water depth of 1.5 m. The sandy pond bottom should be preferred. To prevent the crabs from escaping by burrowing through the bunds, the top width of the bunds should have a minimum of 1.0 m. Also, the overhanging fences on dikes should be fixed as crabs are capable of climbing over the bunds. The height should be between 0.5 to 1.0 m over the dike prepared by materials like bamboo sticks, bamboo poles and knotless nets, asbestos sheets, fiberglass panels *etc.* The 'refugee cages' made up of hollow bamboo pieces, cement pipes or stones are placed inside the pond to minimize mortality and cannibalism on freshly moulted crabs.

The rearing period for soft-shelled crabs is few weeks till their exoskeleton gets hardened. These 'hard' crabs,which are locally known as "mud" (meat) can fetch three to four times better price than the soft crabs.

a. Fattening in Ponds

Firstly the pond bottom is prepared by draining the pond water, sun-drying and adding a sufficient quantity of lime before stocking the soft crabs in the pond. The strengthening of the pond bunds without any holes and crevices should be done carefully. Special attention or care must be given in the sluice area because they tend to escape through nearby areas of the sluice gate. The inlet areas should be protected with bamboo matting inside the bund. The ponds are fenced properly with bamboo poles and nets all along the sides of the bund, which should be inclined towards the pond and prevent the escape of crabs. Soft crabs can be collected from local fishers/crab merchants and stocking should be done in the morning hours at the density of 1-2 crabs/m^2 according to the size of the crabs. The market demand is more for crabs of 550 g and above.

Hence it is best to stock crabs of this size group and the stocking density should not be more than 1 crab/m^2. In this case, 6-8 cycles of "fattening" can be taken out by repetitive stocking and harvesting depending upon the location and availability of water and crabs. If the size of the culture pond is large, then divide the pond into different compartments of suitable sizes and uniform size crabs should be done in the same compartment. This promotes better manipulation of feeding and easy for monitoring and harvest. When the interval between stockings is more, crabs of similar sizes can be stocked in one compartment. It is very much advantageous to do sex-wise stocking in compartments as it reduces the attacks from more aggressive male crabs. To reduce the mutual attacks and cannibalism, it is better to provide shelters like old tires, bamboo baskets, tiles *etc.*

b. Fattening in Cages and Pens

Fattening can be carried out in floating net cages, bamboo cages and pens in shallow estuarine waterways and inside large shrimp ponds having pleasant tidal water influx. The netting material like HDPE, netlon or bamboo splits can be used. In order to ease the feeding and monitoring, the cages of size 3 m x 2 m x 1 m should be arranged row-wise. A recommended stocking density is 10 crabs/m^2 in cages and 5 crabs/m^2 for pens. The tips of the chelae must be removed while stocking to minimize the mutual attacks as the stocking rate is higher in cages. However, like 'fattening' in ponds,these methods are not commercialized.

Grow-out Culture

The grow-out culture is 5 to 6 months in which young crabs are grown to attain a desirable size. Generally, pond based grow-out systems with or without mangroves can be followed for mud crab culture. The pond of size 0.5-2 ha with proper bunds and tidal water exchange should be used. Fencing is advisable in small ponds. In large ponds, strengthening is necessary along the outlet area,especially where natural conditions are prevailing. For stocking, 10-100 g size wild-collected juvenile crabs are used. The culture duration varies between 3-6 months. The stocking density is generally between 1-3 crabs/m^2 provided with supplementary feeding. Usually, feeding should be done with trash fish (wet weight feeding rate - 5 per cent of the biomass per day) in addition to other locally available items. To check the growth and general health and to adjust the feeding rate, regular sampling is necessary. The marketable sized crabs can be partially harvested from the 3rd month onwards. This process of 'stock-thinning' can provide chances for better survival by reducing the mutual attacks and cannibalism.

Feeding

Crabs are fed daily with brackish water clams,trash fishes or boiled chicken waste at the rate of 5-8 per cent of their body weight. If the feed is given twice a day, the major portion of feed must be given during evening hours. The natural diet of mud crab mainly includes crustaceans and mollusks, whereas finfish left overs are found to be very scarce. This is because of their inefficiency to prey upon the fast moving animals. In the grow-out culture management, locally available cheap protein sources (trash fish, mollusks) at the rate of 8-10 per cent of biomass can be given. The crabs can be fed with mixed diet of 25 per cent fish bycatch (trash fish) and 75 per cent fresh flesh of Mollusca or crustacea. Crab biomass can be estimated as the product of man's body weight of stocks in the enclosure and percentage survival.

Harvest and Post-harvest of Mud Crabs

The culture period is generally 3 to 6 months and is determined mainly by the size at stocking and the preference and demand, existing in the market. The culture period may be restricted to 60 days if the crabs having a size of about 250 gm are preferred in the market. Culture duration can be 150 days for *S. tranquebarica* from an initial size of 25 g to a harvestable size of 350 - 450 g if the stocking density is 1 crab per m^2. To obtain a harvestable size of 800-1000g the culture has to be extended further up to 7 months. For *Scylla serrata*, culture duration will be 120 days with an initial size of 25 g and a harvestable size of 200-300 g if the stocking density will be 1 crab per m^2. For obtaining larger sizes upto 400-500 g, the culture period can be extended to further period of 3 months.

Harvest of crabs in a tide-fed pond can be effectively done by letting water through the sluice gate into the pond during an incoming tide. Mud crabs tend to swim against the incoming water and assemble near the sluice gate as the water flushes in and they can be caught with the help of a scoop net. Baited lift nets and bamboo cages/traps can be used for partial harvest. Crabs are to be handpicked after completely draining the culture pond to have a total and complete harvest. Crabs should be tied immediately after their capture in order to curb their movement and to avoid the fighting among themselves and thereby losing their legs. Tying is a process in which a nylon/jute thread is placed in between the frontal portion of the body and the chelipeds and is coiled around their fingers after keeping the chelipeds in folding posture and subsequently double knot made at the rear end of the crab by using both ends of the thread. The "water crabs"caught in the final harvest can be used for fattening purposes. The tied-up crabs are to be initially washed with fresh seawater and subsequently sent for local marketing after packing them in bamboo baskets, in which, they are kept in layers alternatively with materials such as wet seaweeds or moist wood shavings or cotton soaked with seawater to keep the crabs in a cool and moist condition. Those crabs exported in live condition are given a fresh seawater dip and packed in perforated thermocol boxes for air shipment. The expected survival rate during culture would be around 70 to 80 per cent.

Mud crabs are generally sold in live conditions for both local consumption and live crab export trade. For marketing, the mud crabs are graded as "extra-large" (1 kg and above), "large" (500 g to less than 1 kg), "medium" (300 g to less than 500 g) and "small" (200 g to less than 300 g). The female crabs with fully developed ovary are usually sold for a higher price. Live and meaty mudcrabs weighing above 300 g are considered for export, while the undersized live crabs (less than100 g) and those live crabs which have lost their legs can be sold in local markets. While marketing, about 20 per cent of mortality is observed when the transport is by sea, whereas transport by air reduces the mortality to about 5 to 10 per cent. Packing in ventilated and insulated containers instead of cardboard boxes, with 95 per cent relative humidity and 16 – 20°C temperature, will reduce the mortality of the mud crabs during transit up to 7 days and thereby reduce the mortality during transport.

References

1. Balasubramanian, C.P., Cubelio, S.S., Mohanlal, D.L., Ponniah, A.G., Kumar, R., Bineesh, K.K., Ravichandran, P., Gopalakrishnan, A., Mandal, A. and Jena, J.K., 2016. DNA sequence information resolves taxonomic ambiguity of the common mud crab species (Genus *Scylla*) in Indian waters. Mitochondrial DNA Part A, 27(1), pp.270-275.

2. Baliao, D.D., de los Santos, M.A. and Franco, N.M., 1999.Mudcrab, Scylla spp, production in brackishwater ponds.Aquaculture Department, Southeast Asian Fisheries Development Center.

3. CIBA, 2009. Training manual on mud crab breeding and culture.Central Institute of Brackishwater Aquaculture, Chennai, Tamil Nadu.Special Publication No. 45.

4. Cultured Aquatic Species Information Programme.*Scylla serrata*.Cultured Aquatic Species Information Programme. Text by Quinitio, E.T. - SEAFDEC. In: *FAO Fisheries Division* [online]. Rome.Updated 2010.

5. Keenan, C.P., 1999, April. Aquaculture of the mud crab, genus Scylla-past, present and future.In Aciar Proceedings (pp. 9-13).Australian centre for international agricultural.
6. Madhu, K.,andMadhu, R., 2009. CMFRI,Recent Advances in Breeding and Larviculture of Marine Finfish and Shellfish,116-125.
7. Mente, Elena. ed., 2008. Reproductive biology of crustaceans: case studies of decapod crustaceans. CRC Press.
8. Pillai, V. N. and Menon, N. G.,2000, CMFRI,Marine Fisheries Research and Management, *Marichamy R and S. Rajapackiam, paper on Mud crab hatchery technology,747-758.*
9. Raj, T. S., Yohannan, C., Mandal, A., Kumaran, G., Sethuramalingam, A., Srinivasan, P., and Kumar, J. (2015). Status of mud crab industry in India. In Proceedings of the International Seminar-Workshop on Mud Crab Aquaculture and Fisheries Management, 10-12 April 2013, Tamil Nadu, India (pp. 17-26). Rajiv Gandhi Centre for Aquaculture (MPEDA).
10. Shelley, C. and Lovatelli, A., 2011. Mud crab aquaculture: a practical manual. FAO Fisheries and Aquaculture Technical Paper No. 567, FAO, Rome, Italy, pp.
11. Thampi Sam Raj, Y.C., Mandal, A., Kumaran, G., Sethuramalingam, A., Srinivasan, P. and Kumar, J., 2015. Status of mud crab industry in India. In Proceedings of the International Seminar-Workshop on Mud Crab Aquaculture and Fisheries Management, 10-12 April 2013, Tamil Nadu, India (pp. 17-26). Rajiv Gandhi Centre for Aquaculture (MPEDA).
12. Thomas, P. C.,2003. Breeding and seed production of finfishes and shellfishes, page no.246-256.
13. www.fao.org. Biology and culture of selected species.

Chapter 6

Farming of Asian Seabass (*Lates calcarifer*) in India: Present Status and Way Forward

M. Kailasam[1], Pankaj A. Patil[2], Tanveer Hussain[2], Dani Thomas[1], Aritra Bera[1], R. Subburaj[1], G. Thiagarajan[1], and D. Raja Babu[1]

[1]ICAR-Central Institute of Brackishwater Aquaculture, 75-Santhome High Road, Chennai-28
[2]NGRC of ICAR-Central Institute of Brackishwater Aquaculture, Navasari, Gujarat

Introduction

Asian Sea bass (*Lates calcarifer*) known as Bhetki in many parts of India is an important food fish not only in India but also in the Indo-Pacific countries (Thirunavukkarasu *et al.*, 2000: Barlow *et al.*, 1996). Though it forms negligible in perch fishery in commercial catches, it is highly suitable for farming. Sea bass is a euryhaline species capable of withstanding wide salinity fluctuations. It is a catadromous fish migrating towards sea for breeding. Seabass is sexually protoandrous hermaphrodite fish. In the early life stages (2.0 – 3.5 kg body weight) majority of the seabass are males but when they attain the body weight of 4 kg and above, majority become females. Seabass is widely distributed in the tropical Indo-pacific region. It is a fast growing fish and suitable for culture both in earthen ponds and in cage system. Seabass is carnivorous in nature. However, juveniles are omnivores (Barlow *et al.*, 1993). Seabass is an opportunistic predator, whose diet changes at different ages of various size groups. It feeds mainly on zooplankton in early stages and as they grow changes it's feeding on young fishes and shrimps. They show preference for pelagic fish rather than benthic crustaceans as the prey is large. However, juvenile seabass are highly cannibalistic even consume smaller sizes of seabass of the same age group as whole.

Taxonomy

Phylum - Chordata

Sub-phylum - Vertebrata

Class - Pisces

Sub-class - Teleosteii

Order - Perciformes

Family - Centropomidae

Genus - *Lates*

Species - *Lates calcarifer* (Bloch)

Vernacular Names

Tamil - Koduva meen

Malayalam - Kaalanji/Narimeen

Telugu - Pandu kappa

Marathi - Jitada

Bengali - Bhetki

Figure 6.1. Seabass *Lates calcarifer*

Seabass can withstand wide range of salinity from freshwater to 50 ppt with optimal range from 0-36 ppt, which is highly suitable for farming in freshwater, brackish and marine waters. It is a preferred food fish in many south-east Asian countries because of its white tender meat, desired by the consumers. Though, it is opportunistic predatory fish, it can accept formulated pellet feed under culture conditions. High fecundity and relatively simple hatchery production technology make seabass an ideal candidate species for farming in many countries in south East Asian countries and Australia. Seabass can attain the growth up to 1000 g in 8-10 months period either in pond or in the cages. It is highly priced fish both in domestic and international markets depending upon the size. Usually, as the size of the fish increases, the price also increases in the domestic markets.

Species Diversification–Need of the Hour

Farming of finfish species is one of the best choices for species diversification, since they suit for different culture systems such as in earthern ponds, net cages, pens, where wide range of salinity prevail by following either traditional or scientific methods. Amongst the finfish species, the Asian seabass *Lates calcarifer* is highly preferred for farming because of owing to its faster growth rate and high market price. In India, seabass is cultured in ponds traditionally as an extensive type culture wherever seabass is distributed in the wild. Low lying areas and excavated ponds existed in such areas serve as nursery grounds for seabass juveniles and act as wild seed collection centers. Seasonal availability of wild seeds varies with region to region say, April June in West Bengal, May-August in Andhra Pradesh, Sept-Nov. in Tamil Nadu, May to July in Kerala and June-July in Maharastra. Juveniles of assorted size seabass are collected and introduced into the traditional ponds which will be already with some species of fish, shrimps and prawns. These ponds will have the water source from adjoining brackishwater or freshwater canals, or from monsoon flood. The juvenile seabass introduced in the pond will prey upon the available fish or shrimp juveniles as much as available and grow. However, in the recent years, seabass farming is undertaken by obtaining the hatchery produced seed either from CIBA or RGCA. CIBA has developed comprehensive technology for seed production of Asian seabass *Lates calcarifer* in India which earned hope among the farmers for seed availability instead of depending from the wild, which is not only seasonal but also scarce.

Seed Production Technology

Broodstock Development

Fishes weighing 2 – 7 kg size are collected either from wild catches or from the culture system and carefully transported, conditioned and after prophylactic treatment, released into the broodstock holding facilities and maintained. Seabass attains maturity in 3 – 4 years. Females attain maturity when they grow more than 3 kgs and males are in 2 – 2.5 kgs. This size group fishes are maintained in the broodstock holding facilities. Though, the broodstock fishes can be maintained in cages, ponds and concrete tanks, land based concrete tanks are preferred due to various technical advantages.

Concrete Tanks

Broodstock tanks can be 100 to 200 ton capacity. Broodstock tanks should have adequate water inlet and drainage provisions. Flow through facility is desirable and provision for aeration is advisable. The concrete tanks are to be covered with shade nets to prevent the direct sunlight exposure to the tanks which facilitate the algal bloom. Water can be drawn from the open Sea. The desirable range of some of the water quality parameters in the broodstock tank are

Temperature - 28 – 32°C

Salinity - 28 – 35 ppt

pH - 7.0 to 8.2

Dissolved oxygen - more than 5 ppm

Ammonia - less than 0.1 ppm

Nitrite-N - less than 0.01 ppm

Stocking Density

Fishes are maintained @ 1 kg/m^3 in the broodstock tank. Maintaining the fish with quality feed containing good nutritional value is essential for early maturation and spawning. Seabass is a voracious carnivorous fish which feeds mainly on crustaceans/small fishes in live condition. Brood fishes are fed either with fresh or frozen trash fish like Tilapia, Sardines, Horse mackerel *etc.* at the rate of 5 per cent of the body weight. As squid is rich source of protein, it can be given to brood fish atleast once in a week. Enrichment of broodstock with nutritious feed helps in production of good quality eggs which reflects in fertilization rate, hatching rate and larval quality. As prophylactic measure, brood fishes have to be examined every month to assess the health and parasitic infection if any. Depending upon the infection, fishes have to be treated to get rid of infection. The broodstock tanks have to be disinfected periodically with chlorine to avoid cross contamination with parasite and bacteria.

Assessment of Maturation

Maturing fishes can also be implanted with hormone pellet for accelerating the maturation. Besides, maintaining good water quality and health will facilitate to obtain gravid fishes in 6-8 months of maintenance. Matured fishes are selected for induction of spawning. Matured female fishes will have ova with diameter more than 450 μ. Males will ooze milt if the abdomen is gently pressed. The gonadal condition is assessed by ovarian biopsy. Female fishes will be around 4 – 7 kgs and males will be 2.0 – 3.0 kgs. Since sea bass spawning is found to lie with lunar periodicity, days of new moon or full moon or one or two days prior or after these days are preferred for inducing the spawning.

Hormone Dose

After selecting the gravid fishes, the required dose of hormone to be injected is assessed. The dosage level has been standardized as LHRHa @ 60 – 70 μg/kg body weight for females and 30 – 35 μg/kg body weight for males. The hormone can be administered intramuscularly. As the spawning normally occurs in the late evening hours, hormone is injected normally in the early hours of the day between 07.00 – 08.00 hours since the environmental temperature is cool. To ensure proper fertilization normally two males are introduced for one female in the spawning tank.

Spawning

Fishes injected with LHRH-a hormone response for spawning after 30 – 36 hours of injection. Prior to spawning gradual swelling of the abdomen will be seen indicating the ovulation process. Spawning normally occurs late in the evening 19.00 – 20.00 hours. As spawning activity in seabass coincides with lunar periodicity, the activity is found to be in peak during full moon or new moon days. Hence, induced spawning is done during new moon/full moon or one or two days prior or after these days. Seabass is a protracted intermittent spawner (releasing eggs batch by batch) and has high fecundity. In one spawning the fish may release around 1.0 million eggs. The process of spawning will follow during subsequent day also.

Fertilization

Fertilization is external. In natural spawning of seabass in good maturity condition, fertilization will be 70 – 90 per cent. The size of the fertilized eggs will be around 0.75 – 0.80 mm. The fertilized eggs will be floating on the surface and will be transparent. The unfertilized eggs will be opaque and slowly sink to bottom.

Egg Collection, Incubation and Hatching

Since fertilized eggs will be floating on the surface, a bolting net cloth of 150 – 200 µ mesh size can be used for collecting the eggs from the surface. The cloth is stretched as net and towed along the water surface. The collected eggs after washing are transferred to the incubation tanks. The hatching incubation tanks can be 200 – 250 litres capacity and cylindro-conical in shape. Eggs are laid in the hatching tank at the rate of 100 - 200 Nos./litre. Continuous aeration is provided. Temperature of 27° – 29° C is desirable. The eggs will hatch out in 17 – 18 hours after fertilization. The size of newly hatched larvae size would vary from 1.65 to 1.75 µm. After hatching, the larvae can be transferred into larval rearing tanks. The unhatched/unfertilized eggs (dead eggs) in the incubation tank will be removed by siphoning. The larvae can be scooped gently using scoop net and transferred into buckets of known volume. After taking random sample, counting of larvae is carried out to stock the rearing tanks with required stocking density.

Larval Rearing

Freshly hatched healthy larvae can be stocked initially @ 10 – 20 Nos./litre in the larval rearing tanks. Depending upon the age and size, the larval density can be reduced to 10-15 Nos./l on 10th day and after 15 days, the density is maintained around 8-10 Nos./l. Supply of appropriate live feeds with required nutrients to the seabass larvae is an important step to achieve better survival rate (Rimmer *et al.*, 1994). Rotifer (*Brachionus plicatilis*) can be given as initial feed to the larvae from 3rd day post hatch (Kailasam *et al.*, 2007). Rotifer is maintained in the larval rearing tanks at the rate of 2-4 Nos./ml initially. From 4th day to 15th day the rotifer concentration is increased to 6 – 10 Nos./ml. Every day after water exchange, the food concentration in the tank should be assessed and fresh rotifers should be added to the required concentration (Dhert *et al.*, 1990). *Artemia* nauplii are given as feed along with rotifers and green water from 10th day. By this time, the larvae will reach to the size of 4 mm total length. Larvae can be fed exclusively with *Artemia* from 16th day to 21st day post hatch. The density of the brine shrimp nauplii in the rearing medium can be maintained @ 2-4 Nos./ml on 10th Day Post Hatch (DPH) and gradually increased to 6-10 Nos./ml up to 21 DPH. The daily ration of *Artemia* nauplii feeding is adjusted after assessing the unfed *Artemia* in the rearing tank at the time of water exchange and the larval density. Meanwhile, formulated larval diet with the particle size of 100-200µm can be also provided to the larvae from 17 DPH and slowly it can be weaned to the artificial diet. The size of the particulate feed can be increased to 500µm by 21st DPH and by maintaining these protocols, the seabass larvae can be totally weaned to the artificial diet by 30th DPH. After this day, the larvae will attain the fry size 1.8 to 2.0 mm at 45 DPH. The seabass fry is ready to supply to the farmers for nursery rearing.

Water Change

Water quality in the rearing tanks is very important for better survival and growth of the larvae. Water provided to the larval rearing tanks should be free from flagellates, ciliates and other unwanted pathogenic organisms. Water should be filtered through biological filters, pressure sand filters and treated with UV radiation, to get rid of the pathogenic organisms. If chlorine treated water is drawn, residual chlorine should be removed, since, fish larvae are highly sensitive to chlorine and water should be used only after de-chlorination.

Grading

Seabass while growing exhibits differential growth rate, hierarchy, resulting different size groups in the same rearing tank. The large one's called 'shooters' dominate others for food and space and also prey on them. In the rearing tanks, when the larval concentration is more and congregation takes place for food and feeding, the larger ones are tempted to feed on the smaller ones. To avoid this problem, regular grading has to be done by removing the shooters. Uniform sized larvae should be maintained in the rearing tanks for better survival and growth (Kailasam *et al.*, 2002). Grading should be done once in four days from 15th day or whenever different size larvae are seen in the tanks. Grading can be done using a series of fish graders with different pore size of 2 mm, 4 mm, 6 mm, 8 mm and 10 mm. When the larvae are allowed to pass through the graders, different size will be retained according to pore size of the sieves. Grading may cause injuries which lead to mortality. Hence proper care should be taken in handling the larvae.

Nursery Rearing

Seabass nursery rearing is done in three different systems such as tank based, hapa net based and pond based depending upon the resource available

Nursery Rearing in Tanks

Seabass fry in the size of 1.8–2.0 cm can be stocked at the rate of 5 Nos./l in circular or rectangular (RCC or FRP) nursery tanks with 5-10 ton capacity. Semi-outdoor tanks are preferable for nursery rearing. The tanks should be provided with inlet and outlet provision for easy filling and draining of water. RAS (Recirculation Aquaculture System) or flow through system is desirable for obtaining good growth of fries. Proper feeding and maintenance of water quality are highly essential activities in the nursery phase. Artificial feed in the size of 500-800 µm can be supplied in 3-4 rations @ 8-10 per cent of body weight initially and later 6-8 per cent. Unfed feed have to be removed everyday and grading has to be done weekly twice depending upon the shooters emergence. The nursery rearing tanks have to be disinfected at the interval of 3-4 weeks using bleaching solutions as per the requirement. In the flow through system, water exchange is to be carried out daily to an extent of 100-200 per cent. After 60-75 days of rearing, the fry can attain the stage of fingerling having the size of 3-4 inch with survival rate of 70-80 per cent.

Nursery Rearing in Ponds

Size of the Nursery ponds will be around 200-250 m^2 with provision of retaining water to a level of 70 – 80 cm. Adequate provision for water inlet and water drainage should be provided. Gentle slope is provided from inlet to outlet to let out the water by gravity. Suitable sized (normally 1 mm) mesh screen nets should be provided in the inlet side and outlet side to avoid entry of unwanted fishes and escape of the stocked fish respectively.

Figure 6.2. Nursery Rearing in Ponds.

The pond is prepared before stocking. If there are any predator/pest fishes they have to be removed by repeated netting, draining and drying the pond. In case where complete draining is not possible, water level is reduced to the extent possible and treated with Derris root powder @ 20 kg/ha added or mahua oil cake @ 200-300 kg to eradicate unwanted fishes. After preparing the pond, good quality water is filled to a limited depth. If the pond bottom is acidic, neutralization is done with lime application. In order to make the natural food abundant, the pond is fertilized with cow dung manure @ 500 kg/ha keeping the pond water level 40-50 cm. The water level is gradually increased. After 2-3 weeks period when the natural algal food is developed well, freshly hatched *Artemia* nauplii are introduced. Normally 1 kg of cyst is used for 0.5 ha pond. These stocked nauplii grow and become biomass in the pond forming food for the seabass fry. Small size seabass fry (0.8-1.2mm) is stocked @ 20-30 Nos./m^2 in the early hours of the day. Fry should be acclimatized to the pond condition before stocking.

Supplementary feeding is done with artificial pellet/powdered feed in a gradual manner to wean the seed to the artificial diet. The feeding rate can be @ 10-12 per cent body weight initially and later reduced to 6-8 per cent. Excessive feeding should be avoided since it would deteriorate the water quality and pond bottom. Excessive algae if any found in the pond should be removed. In a rearing period of 6-8 weeks, the seed will attain in the size of 7-10 cm with survival rate of around 60 per cent.

Nursery Rearing in Cages/Hapas

Rearing the seabass fry in cages/hapas is commonly practiced. This method is advantageous than other methods since the management is easy and requires less space and capital investment for installation of rearing facility.

Figure 6.3. Nursery Rearing in Cages/Hapas.

The size of the hapas will be chosen either as 2 X 1 X 1 or 2 X 2 X 1 m depending on the requirement. Hapas are made with nylon/polyethylene webbing with mesh size of 1.5-3.0 mm. The stocking size Seabass fry is 1.8-2.0 cm and stocked @ 500 Nos./m^2 (Kailasam *et al.*, 2006). The fries are to be fed with specialized formulated farm made nursery feed during nursery rearing (Biswas etal., 2010). Feeding strategies are similar as described in the nursery rearing in tanks. The net cages have to be checked daily for damages caused by crabs and others and clogging by the adherence of suspended and detritus materials to avoid restriction in the water flow. It could be avoided by cleaning the cages or hapas every day. Regular grading (twice a week) should be done to avoid cannibalism and increase the survival rate (Krishna Sukumaran *et al.*, 2011). In this method of rearing, farmer can get survival of seabass fries upto 65-80 per cent. Seabass nursery rearing activity for fingerling production is popularizing among the farmers and self-help groups, due to high demand of fingerling seed for grow out culture.

Pre-Grow Out Culture Technology

Pre-grow out culture is the stage between nursery and grow out culture, in which the 5-10 g size fingerling seabass is grown to 80-100 g stockable size for grow out culture for the period from 60-75 days culture under pond condition with the expected survival rate from 45-60 per cent. In this rearing, seabass fingerlings are stocked at the rate of 10000-15000 Nos./ha in small size ponds in the size of 0.1 to

0.2 ha since small pond facilitates easy operation. Similarly, small volume cages in the size of 2 x 2 x 1m with the mesh size of 16-18 mm are also preferred. In the pre-grow out cages, stocking can be done at the rate of 30-40 Nos./m^3 and feeding will be done at the rate of 8-12 per cent and 6-8 per cent body weight in the pond and cages respectively. Total fish biomass should be measured fortnightly by taking random sampling of fish and accordingly feed quantity will be decided. Care must be taken to maintain the optimal feed quantity while feeding daily since under feeding may cause differential growth which may leads to cannibalism and reduction in the survival rate. Regular grading has to be done to maintain the uniform size of seeds.

Grow Out Culture Technology

Traditional Culture

Seabass is cultured in ponds traditionally as an extensive type culture by the farmers. In low lying excavated ponds, whenever the seabass juveniles are available from the wild, seabass are allowed to grow for 6-7 months or till such time water level is available in these ponds and then harvested. In this manner, production from 0.5 to 1.5 ton/ha can be achieved. However, this practice is highly unorganized and uncertainty in production or return for the farmers. With the advances in the technology in the production of seed under captivity assuring the supply of uniform sized seed for stocking and quality feed for feeding, the seabass culture can be done in more organized manner.

Improved Seabass Culture Methods

In this method, stocking of uniform sized seed at specific density and feeding with low cost formulated feed of required quantity seabass farming can be practiced. Water quality is maintained with periodical exchange of water. Fishes are allowed to grow to marketable size, harvested and marketed for high unit price. Seabass culture can be done in more organized manner as a small-scale/large scale aquaculture in brackishwater and freshwater ponds in cages. Seabass is cultured in the ponds adopting poly culture method or feeding with low cost fishes like Tilapia/oil sardines or with extruded floating pellets.

Grow Out Culture of Seabass in Cages

Fish culture in cages has been identified as one of the eco-friendly at the same time intensive culture practice for increasing the fish production. Cages can be installed in open sea or coastal backwaters. Cage culture system allows high stocking density, assures high survival rate. In cage culture, management of feeding, disease and cage can be done easily except the water currents and water quality parameters. Fishes in the cages can be harvested as per the requirement of the consumers, which will fetch high unit price. Cages can be relocated whenever necessary to avoid any unfavorable condition. Grow out cages in the size of 10 to 50 m^2 are preferable for easy management and maintenance. Cages are fabricated with polyethylene netting with mesh size ranging from 18 to 24 mm depending upon the fish size to be stocked. There are two types of cages namely floating and stationary cage.

Floating Net Cages

The net cages are attached to wooden or GI frames kept afloat using plastic drums. Concrete weight blocks as anchors can be attached to the corners of the net cage at the bottom. These types of cages can be installed in areas with water depth more than 3 meters with feeble water current.

Stationary Net Cages

These are fixed enclosures, which can be installed in shallow water areas in lagoons, brackishwater lakes having water depth of 2-4 meters. The cage net is fastened to wooden poles erected in the water system at the four corners.

Stocking Density

In the cages, fishes can be stocked at the rate of 25-30 Nos./m^2 initially when they are in the size of 80-100 g. Fishes in the cage can be fed with wither slow sinking or floating pellet feed. The rate of feeding can be maintained at the rate of 5-6 per cent initially and later 3 per cent body weight. After 6-8 months of culture period, the fish will attain the average body weight of 1.0 kg with the productivity of 2-25kg/m^3. Grading and cage management has to be carried out regularly to get better survival rate and higher productivity in the cage system.

Pond Based Culture System

Earthern pond in the size of 0.5 to 1.0 ha is preferred for grow out culture. Preparation of the pond can be done by drying, tilling, levelling initially and seawater filling can be done up to 50-60 cm depth. The pond can be manured with raw cow dung @1000 kg/ha. In addition, fertilizers like urea @100kg/ha, super phosphate @ 50kg/ha can also be added to enhance the algal bloom in the pond. Fishes like Tilapia can be introduced @ 20-25 thousand/ha into the pond. Stocking of forage fishes should be done in the seabass culture pond a month ahead of seabass stocking. The water level in the pond may be maintained around 100-120 cm. Subsistence level of feeding with cheaper ingredients like Rice bran and ground nut oil cake @ 2-3 per cent of the biomass can be done. Being a prolific breeder in habit, the forage fish *Tilapia* will start breeding in the pond and at given point of time in the proposed seabass culture pond, there will be different size groups of Tilapia will be available. In the pond with natural forage feed, seabass of 80-100 gm can be stocked at the rate of 4000 - 4500/ha depending upon the management capacity of the farmers like water exchange and feeding at later stages. Seabass should be uniform in size at the time of stocking to avoid differential growth in the culture pond. The Seabass stocked will feed upon the hatchlings fry and fingerlings of the forage fish initially and later on the adult fishes also. Since, seabass has got higher growth potential than the forage fishes they will dominate in the pond. After 3-4 months there will not be any forage fish in the pond. At that stage, fresh stock of feed fishes should be introduced. In this trash fish feeding method, seabass production can be achieved @ 3.5 to 4.0 Ton/ha in 6 to 8 months culture period with the trash fish FCR in the range of 6.0 to 7.0. Seabass farming with trash fish feeding has limitations since, trash fish is expensive in many places and availability is scarce. Therefore, it is recommended that seabass farming can be done with specially prepared pellet feed for mass scale adoption of this technology.

Figure 6.4. Pond Based Culture System.

Way Forward

Any product that especially from agriculture produce has to build strong marketing network for smooth flow of movement in different segment of clients with addition of value. These chains of network activities construct a strong demand and supply process in the market with value addition for particular product. The value chain network surge usually in two dimension which are vertical dimension and horizontal dimension

The Vertical Dimension Reflects the Flow of Product from Primary Producer to End-consumer

In seabass farming, seed and feed are the two important inputs that require at all level till reach to final consumer. A range of products such as seabass fry, fingerlings, stockable size juvenile (50-100 g size) and marketing size harvested seabass (above 1.0 kg) are in demand among the actors involved in the chain link. The flow of products determined by the factors such as quality of the seed (size and quantity available), cost of the seed, technology available for packing and transportation for different products, price, location of the hatchery or farm for easy accessibility to reach by road/rail/air for smooth transportation. Similarly, availability of larval diet, nursery feed and grow-out feed either in floating or slow sinking/sinking pellets in adequate quantity at the farming site are other deciding factors in the network chain of seabass farming.

The Horizontal Dimension Reflects Relationships between the Stakeholders in the same Chain Link (between farmers, between processors, *etc.*)

The linkages between the hatchery operators, nursery rearing farmers/SHGs, pre-grow-out culturists, grow-out culture farmers, and marketing agents for lifting the produce from the farms, vendors of local markets, processors and exporters for international markets are important factors which would decide the smooth flow of finished products to the end consumers

Movement of Seabass Seed Supplied to the Farmers in different Routes and Destinations

Seabass seed can be transported either by road or by rail or even by air. When small size seed can be packed in oxygenated polythene covers and transported, bigger size seed can be transported in the open container with oxygen. The density can be decided based the distance and transportation time. Spawned eggs, larvae and early fry are usually packed reasonably in good density and advanced fry and fingerlings are usually transported in the open container for long distance and if within 12 hrs, the seed is packed in low density.

Marketing Sacilities

Seabass has very good demand in the local market especially in major cities. However, the price varies from place to place. Higher price prevail in some season especially during festival season when compare to other months. Continuous production and supply, cold storage facility and, transportation facility are some of the deciding factors of the price in the farm site.

Conclusion

The demand for the seabass seed among the farmers for the culture is increasing every year. It is approximately estimated that the demand for the seabass fry in India would be 100 million/year. Presently, a total of around 5 million seabass fry is produced from two government hatcheries and could hardly meet 5 per cent of the total demand every year. Therefore, it is high time that private hatcheries have to be established in different parts of the country to meet the seed demand. Since, hatchery establishment is capital intensive, financial institutions and promotional bodies can support by providing required loan/subsidiary to the hatchery operators. On the other hand, establishment of satellite seabass seed rearing facility in different locations would also help the farmers to source the stock size seed at local level to expand the seabass farming.

References

1. Barlow, C.G., Rodgers, L.J., Palmer, P.J., Longhurst, C.G., 1993. Feeding habits of hatchery-reared barramundi *Lates calcarifer* (Bloch) fry. Aquaculture 109, 131-144.

2. Barlow, C., K. Williams and M.Rimmer 1996. Seabass culture in Australia. Infofish International, 2: 23-26

3. Biswas G, A. R. T. Arasu, J. K. Sundaray and M. Kailasam, 2010. Optimization of feeding frequency of Asian seabass, (*Lates calcarifer*) fry reared in net cages under brackishwater environment.

4. Dhert, P. M. N.Duray, P.Lavens and P. Sorgelos. 1990. Optimized feeding strategies in the larviculture of the Asian seabass *Lates calcarifer*. In : R.Hirano and I Hanyo (Eds), Proc. The Second Asian Fisheries Forum, Asian Fisheries Society, Manila, Philippines p 319-323

5. Kailasam. M, A. R. Thirunavukkarasu, Mathew Abraham, P.Kishore Chandra and R. Subburaj. 2002. Influence of size variation and feeding on cannibalism of Asian seabass *Lates calcarifer* (Bloch) during hatchery rearing phase. Indian Journal of Fisheries. 49 (2) 107-113

6. Kailasam, M., Kishore Chandra, P., Muralidhar, M, Thirunavukkarasu, A.R., Mathew Abraham, 2001. Survival and growth of sea bass *Lates calcarifer* (Bloch) fry reared at different stocking densities. In: Perspectives in Mariculture (eds. N.G. Menon and P.P. Pillai), pp. 311-320.

7. Kailasam, M., Thirunavukkarasu, A.R., Sundaray, J.K., Mathew Abraham, Subburaj, R., Thiagarajan, G., Karaiyan, K., 2006. Evaluation of different feeds for nursery rearing of Asian sea bass *Lates calcarifer* (Bloch). Indian J. Fish. 53 (2), 185-190.

8. Kailasam M, Thirunavukkarasu A.R, Selvaraj, S and P.Stalin, 2007. Effect of delayed initial feeding on growth and survival of Asian seabass *Lates calcarifer* (Bloch), Aquaculture 271 (2007) 298-306

9. Krishna Sukumaran, A.R. Thirunavukkarasu, M.Kailasam, J.K.Sundaray, R.Subburaj and G. Thiagarajan. 2011. Effect of stocking density on size heterogeneity and sibling cannibalism in Asian seabass *Lates calcarifer* (Bloch) larvae. Indian J. Fish. 58 (3), 145-147.

10. Rimmer, M.A., Read, A.W., Levitt, M.S., Lisie, A.T., 1994. Effects of nutritional enhancement of live food organisms on growth and survival of Barramundi, Lates calcarifer (Bloch) larvae. Aquacult. Fish. Management. 25, 143-156.

11. Thirunavukkarasu, A.R, M.Kailasam, P.Kishore Chandra, P.Shiranee, Mathew Abraham, A.V.K.Charles and R.Subburaj, 2001. Captive broodstock development and breeding of seabass *Lates calcarifer* (Bloch) in India. In : Perspectives in Mariculture (eds. N.G.Menon and P.P.Pillai) 111-124.

Chapter 7

Seed Production and Culture of Shrimps

C.P. Balasubramanian, P.S. Shyne Anand and K.K. Vijayan

ICAR-Central Institute of Brackishwater Aquaculture, Chennai – 600 028

Indian brackishwater aquaculture, almost exclusively dominated by shrimp, has been the most dynamic and socially emotive food production system. In India, shrimp farming through 1990 to present has become a multibillion dollar industry. In 2018, frozen shrimp earned almost US$ 5 billion, which is about 70 per cent of the total seafood export earnings. Seed production and grow-out production systems are the major component of shrimp farming. These documents provide the guidelines for shrimp hatchery and grow out production system

Seed Production and Management

Seed production is the initial phase of any aquaculture enterprise. Hatcheries are the infrastructure for the mass production of post larvae, which is critical for the development of aquaculture. The first larval rearing technique of shrimp was adopted from the Japanese method that utilizes the large tank system, subsequently this system was modified by using small tanks following the Galveston system. Shrimp seed production sector has been categorized into three phases: post larval phase, hatchery phase and breeding phase. During the early phase of shrimp farming, wild collected post larvae were stocked in aquaculture ponds. Earlier research work carried out by CMFRI paved the way for commercial hatchery production in India. First commercial hatchery production of *Penaeus indicus* was started by Kerala state Fisheries Department at Azhikode and a Private hatchery in

Kerala. Later Hindustan Lever started a shrimp hatchery for production of *Penaeus monodon* PL near Chennai. Through the last three decades from 1990 to 2020 the hatchery production protocol has been refined and shrimp hatchery production industry has become a sophisticated industry.

Initial two decades, from 1990 to 2010, hatchery production in India was exclusively the production of *Penaeus monodon*. Subsequently when shrimp industry shifted from *P. monodon* to exotic *P. vannamei*, most hatcheries in India were converted to *P. vannamei* seed production. Although protocol for mass production of larvae of both species are essentially similar,

Biology of *Penaeus monodon* and *P. vannamei*

Penaeid shrimps are included in the decapod suborder dendrobranhiata, which are distinguished by other shrimp/prawn like crustacea by the difference in gill structure and spawning habit. Penaeid shrimps are characterized by 'tree- like gill' (dendrobranch) and they release the eggs directly to the sea while spawning. About 500 species of shrimps have been reported so far under the dendrobranchiate group.

Nomenclature

Most shrimps used for brackishwater shrimp culture belongs to the genus *Penaeus*. Although this genus split into six valid genera (*Litopenaeus, Fenneropenaeus, Penaeus, Farfantepenaeus, Marsupenaeus* and *Melicertus*) in 1997 by Perez Farfante and Kensley, the subsequent molecular studies found no valid evidence for this revision. Therefore, it is preferred to use old genus name *Penaeus* to describe all the cultivable shrimp species under this taxa.

Life Cycle

Most cultivable species of penaeid shrimp show similar pattern of life cycle. The adults live in the sea and reproductively mature in this environment. They lay eggs in the sea and larval development takes place in this habitat. Post larvae drift towards the coast and enter into the less saline (brackishwater environment) ecosystem such as estuaries, brackishwater creeks and lakes. They grow into juveniles and after spending 4-6 months in this environment, sub adults migrate back to the sea where they grow and attain sexual maturity and complete life cycle. The natural ability of the post larvae and juveniles to live and grow in the fluctuating salinity conditions of estuarine environment has been made use of culture them in brackishwater ponds

Reproduction

Sexes are separate. Males and females can externally be distinguished at smaller size; male possess a small rudimentary petasma, and thelycum of female is discernible as a small elevation between fourth and fifth pereiopods. The males are generally mature at a smaller size than females. Male genital system comprises of paired testes, paired vas deference and paired terminal ampoule. In addition to these internal organs, a paired petasma and paired appendix musculina are found externally. The female reproductive system consists of paired ovaries, paired oviduct and single thelycum.

Based on the reproductive morphology female penaeid shrimp can be classified into two groups: Closed thelycum group and open thelycum group. Females of all shrimps and prawns have a specialized region on the ventral side between third and fifth pereopods/walking legs. It serves as a genital area for genital contact of males where they deposit the spermatophores or sperm sac during mating, and female carries these spermatophores either temporarily or for a prolonged duration. This region is termed as thelycum, the Greek meaning is 'women like', and it is unique in each species and sometimes used as a diagnostic character for species identification. Closed thelycum is a special kind of structure and it is observed in all Asian species of genus *Penaeus viz., P. monodon, P. indicus, P. merguiensis, P. penicillatus,P. japonicus, P. semisulcatus*etc, as well as Pacific and Atlantic species *viz., Penaeus vannamei, P. stylirostris, P. occidentalis, P. schimitii and P. setiferus*)

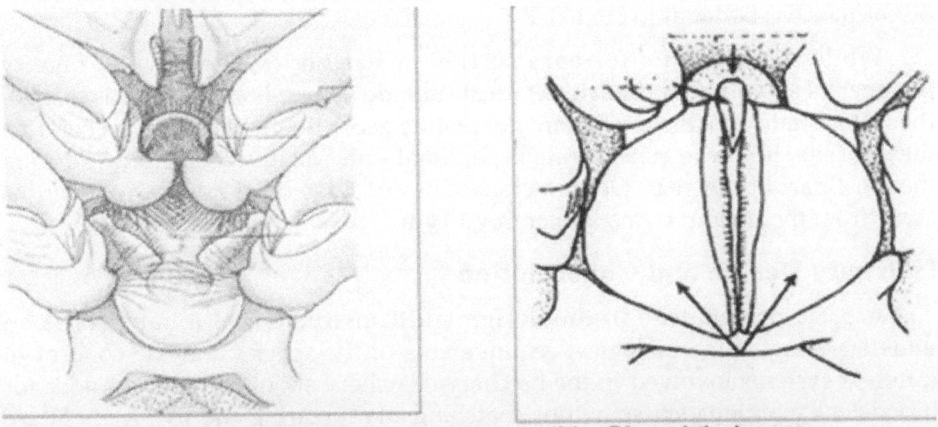

a) Open thelycum b) Closed thelycum

Figure 7.1. Types of Thelycum.

Tiger shrimp, *P. monodon*, is the largest shrimp, native to Indian Ocean and south western Pacific Ocean form Japan to Australia. The maximum size recorded is 362 mm. Once it was the most dominant cultured shrimp, it accounts almost 80 per cent of total farmed shrimp production. Later the farming of this shrimp was replaced by white leg shrimp, *P. Vannamie* which is distributed from Peru to Mexico.

Hatchery Site Evaluation

One of the major factors that determine the success of shrimp hatchery is the proper site evaluation. Apart from that, ecological suitability, climatic suitability and Government guidelines are the major factors also play a role in success of hatchery operation. Ecological suitability includes the parameters like salinity, temperature, pH, DO, organic and sediment loads, absence of heavy metal, and pesticides are the major factors in shrimp hatchery operations.

Salinity is the most important water quality characteristic affects the physiology of penaeid shrimp. In nature, penaeid shrimp migrate to sea during the adulthood, and reproduction occurs during this phase. Therefore, shrimp require oceanic salinity for reproductive maturation and larval production. The recommended salinity level for hatchery production of shrimp is 28 to 36 ppt. Heavy rain fall and freshwater influxes reduces the salinity, and it becomes unsuitable for hatchery operation. Therefore, it is not recommended to construct hatcheries at river mouth, or areas experiences heavy rainfall. However recent refinement in the water management and use of Recirculation aquaculture system (RAS), hatcheries could be constructed away from the sea, by using diluted brine.

Temperature should be maintained between 27 to 29° C. Although tropical species such as *P. monodon* will perform optimally at high temperature, *P. vannamei* needs a lower temperature, for maturation many hatcheries in India use chiller to reduce the temperature. In *P. monodon* hatcheries during winter season titanium heaters are used to maintain the optimum temperature. Maturation and larval performance is better at pH 8 to 8. 2.

While selecting site for construction of hatcheries, these water quality parameters should be conceded. As ideal sites are almost non-existent, as most of the water quality characteristics are fluctuating according to the season. Therefore the ideal site should be a site having the optimal water quality characteristic during the most part of the year. Other factors to be considered are hatchery should be away from the shrimp farms, preferably 10 km.

Hatchery Design and Construction

A basic prerequisite in the design and construction of a hatchery is an understanding of the biological requirements of the species and the concept of different systems involved in the hatchery. Hatchery should have provisions for broodstock maintenance, spawning/hatching larval rearing and live-feed culture (phytoplankton and zooplankton). The maintenance of these biological materials require 4 major infrastructure systems: Seawater supply system, b) Air-supply system, c) Tanks and d) Buildings.

Seawater Intake

There are numerous designs now in use in the various sea water based hatcheries. These designs are dependent on specific site characteristics, topography, geology, climate *etc.* Mostly used system of drawing water is through intertidal borewells or through inshore open wells. Low depth intertidal borewells are suitable in areas where the wave action is minimal. Inshore open wells could be used where the wave action is more in the intertidal zone and there is no freshwater aquifer in the shoreline. Where the water from the inshore wells are Iow saline due to freshwater table, drawing of water is done directly from open sea by constructing concrete jetties into the sea beyond the breaker zone.

Water Quality

Water quality of shrimp hatchery depicts the sum total of physical and chemical characteristics of the oceanic water that support the healthy larval development. A number of steps should be taken for achieving this including filtering through sand filters (gravity and pressure) and settling in reservoir tank followed by disinfection using chlorine and final disinfection by UV or Ozone.

Better Management Practices for Water Quality

Water Treatment System should be Designed to provide High Quality Oceanic Seawater

- Before the intake of water into the hatchery, the salinity of water should be checked. The salinity at the range of 33-34 ppt is optimum, although 29-30 ppt is also acceptable.
- As best salinity and high quality of seawater is found during spring tide, intake water should be taken during this time. Seawater can either be pumped into settlement tank for 1-3 days until all the suspended matter settled at the bottom or passed directly through the sand filters (large gravity flow filters/pressurized sand filters)
- Intake seawater should be disinfected, and disinfection is done by chlorination. Chlorine can either be used as powder (calcium hypochlorite: 60-70 per cent active chlorine) or liquid (sodium hypochlorite: 7-10 per cent active chlorine). An active chlorine level of 10-20 ppm for 12-24 h is sufficient to kill most pathogens. In order to achieve 10 ppm chlorine level 15 g of 65 per cent calcium hypochlorite or 100 ml of 10 per cent sodium hypochlorite should be added to every 1000L of seawater. Aeration should be turned on for 10-15 minute for effective mixing of chlorine, and after having mixed well, aeration should be turned off and allow the water to remain for 10-12 h. If aeration is not turned off, the chlorine concentration in the water dissipate fast and will not be able to kill pathogen. After 12 to 24 h water should be checked for residual chlorine by using commercial chlorine kit, and chlorine should be neutralized with sodium thiosulfate at the rate of 1 ppm for every 1 ppm chlorine concentration.
- It is also advisable to pass chlorinated water through activated carbon filter and further disinfection with UV or ozone. UV irradiation must reach >30000 mw/cm^2 and ozone concentration should be > 0.5 µg/ml for 10 minutes for effective disinfection of virus.
- Each functional unit must receive water that is treated and preferably must have separate water supply systems. If possible, re-circulatory system in specific sections is to be added to ensure water quality with reduced water usage

Hatchery Seawater Intake and Treatment Process

Figure 7.2. Reservoir/Settlement Tank.

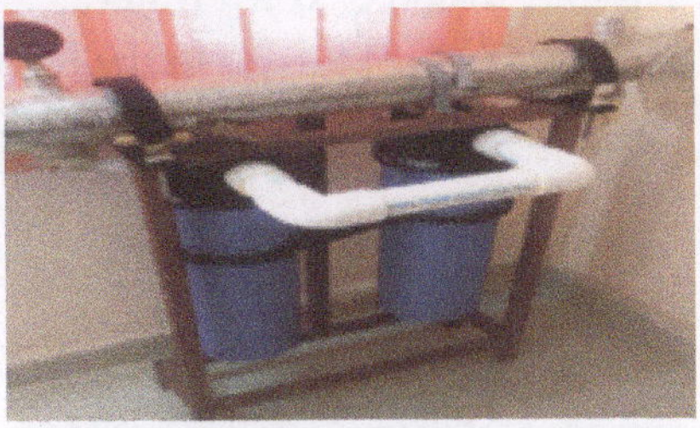

Figure 7.3. Catridge Filder and UV.

Broodstock Management

The use of high quality broodstock and efficient management is the corner stone for the success of hatchery production. In India, both *P. monodon* and *P vannamei* are identified for aquaculture.

Figure 7.4. *Penaeus vannamei.*

Figure 7.5. *Penaeus monodon.*

In the case of *P. vannamei,* hatchery production is based on the imported Specific Pathogen Free (SPF) stock and for *P. monodon,* wild caught broodstocks are used. Broodstock accounts for >30 per cent of production cost of hatchery, and therefore the management of broodstock should be efficient. Normally, in the case of *P. monodon,* live broodstock are collected by trawlers or country boat operators. In case of trawlers, the duration of trawling is more than 2 to 3 hours and this leads to high level of stress to broodstock. In some cases the total fishing period will be for more than 2 to 3 days and the shrimps are kept for long duration in on-board in small containers leading to stress and cross contamination between infected and non-infected stock. The sea water used for transportation of the broodstock in the boat and also at the port collection centre is generally not filtered and likely to contain pollutants and pathogens which lead to contamination of the stock

Better Management Practices (BMP) for Broodstock Management

- ✰ For capture of shrimp broodstock, short duration of trawling should be done.
- ✰ The collection and transportation should be stress free and the fishing ground should be as far as from the coast. The captured stock should be immediately brought to the shore. Collection boats could be used to collect from different trawlers and move the broodstock to the hatchery.
- ✰ The broodstock after capture should be kept individually in filtered and disinfected sea water.
- ✰ To avoid stress to shrimp, they can be kept under oxygen packing in low temperature (18-28°C) and stocking should not be more than 1 number per bag. Placing a rubber tube over the rostrum will stop puncturing the bag.
- ✰ To chelate the heavy metal, EDTA (10ppm) and to control nitrogenous metabolites, 1g/L activated carbon or probiotics may be added to the water

- The hatchery should have a strong wall or peripheral boundaries, vehicle dip, shower and change room and foot dip with disinfectant to reduce any biosecurity risk and contamination of hatchery.
- As far as possible broodstock should be held individually until their disease status for WSSV and MBV/or any emerging pathogen is checked.
- For *P. monodon*, the female broodstock should have a weigh at least 120 g (>28 cm) whereas the males could be at least 70 g (21cm length).
- Upon arrival at the hatchery, the broodstock should be treated with 100 ppm $KMnO_4$ or liquid povidone for 1 min before being stocked in the tanks.
- Important existing pathogens should be screened individually for each broodstock by non-lethal sampling and microscopic observation for the MBV should be done by observing the occlusion bodies in the faeces with malachite green stain/PCR method.
- The testing for all important existing pathogen could be repeated after spawning and if confirmed, the entire batch should be discarded after proper disinfection
- If any of the shrimps start dying after checking negative for virus, hemolymph bacterial count should be estimated in the surviving shrimp and if vibrio is detected in significant numbers in hemolymph, probiotics and vitamin C can be given through feed. However shrimp with black melanized lesions in the body or white muscles are bright red in color should be discarded immediately before these symptoms appear in others.
- For import of SPF broodstock, the hatchery operator should import from the approved suppliers of CAA.
- Only Specific Pathogen Free broodstock cleared through the quarantine should be used in the hatchery for the seed production.
- Use of broodstock other than those certified SPF by the designated authority is strictly prohibited.
- Hatcheries involved in SPF seed production should not use any other non SPF species within the hatchery premises.
- Stocking density in the maturation tanks should be restricted to 2-3 Nos./m^2 for P. monodon and 3-4 Nos./m^2 for P.vannamei
- After holding for a week, only hard-shell (inter molt/late pre molt) females kept in chilled seawater (20-25°C), are ablated by cutting one eye with appropriate method. Pre-moult or post-moult shrimps can be used after a week or so they can withstand the ablation stress
- Feed shrimp with high quality feed including polychaete worm, squid, mussels, enriched adult Artemia and dry broodstock diet
- Feed should be provided at the rate of 25 per cent of wet body weight and should be divided into 6-8 times a day. Before each feeding the uneaten feed is to be removed.

- ✰ Immediately after spawning (usually between 8 pm-2 am), the female should be taken out and placed back into a clean maturation tank. The eggs can be transferred after proper washing to separate egg hatching tanks filled with clean seawater with 20-30 ppm EDTA and 0.05-0.01 ppm treflan
- ✰ Preferably, each female broodstock can be spawned to a maximum of 3 times in case of *P.monodon* and 10-12 times in case of *P. vannamei* before discarding in order to maintain high quality nauplii. Unless the female spawn three times before moulting, it should not be used again. Hatching tank should be illuminated with good light source which increases the hatching rate. Only healthy nauplii, attracted to light should be harvested at around noon of the day following spawning, disinfected and stocked into larval rearing tanks. If a single spawn shows a low hatching rate (<40 per cent), it should be considered unhealthy batch and discarded

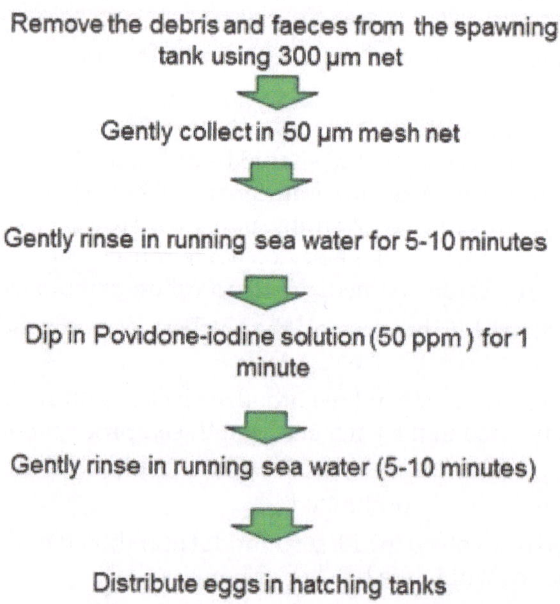

Figure 7.6: Schematic Flow Chart for Washing and Disinfection of Eggs in the Penaeid Hatcheries.

- ✰ Spawning tanks should be placed in dark rooms without disturbance to the spawners and best quality water should be available for spawning and hatching.
- ✰ Spawners should be transferred to the spawning tanks in the late evening and no feeding should be done after that since it will contaminate the eggs at release.
- ✰ Mother shrimp should be removed immediately after spawning.

- ✰ Eggs should be collected, washed thoroughly and disinfected (the washing and disinfection procedure are given in the flow chart) before disbursing in hatching tanks with fresh filtered and disinfected sea water. The eggs should be checked for presence of pathogen. The washing procedure of the eggs is given as flow chart.
- ✰ Hatching tank should be 200-500 L in volume for individual hatching (individual hatching is preferred). It is ideally to be stocked 1 million eggs per 1000 L of water. No aeration or slight aeration should be given until nauplii hatching (hatching takes place after about 8 h). If aeration is not provided to the eggs, periodic agitation has to be given to prevent piling up of eggs at the bottom.
- ✰ After hatching is completed, the active nauplii should be collected using their positive phototactic behaviour.
- ✰ The collected nauplii should be washed and disinfected with formalin or any other known disinfectant before returning them to the tanks with fresh filtered and disinfected sea water. Nauplii should be counted, and if hatching rate is less than 40 to 50 per cent it should be discarded as low hatching rate indicates poor spawn quality. They should be distributed in larval rearing tanks when they reach N5 stage

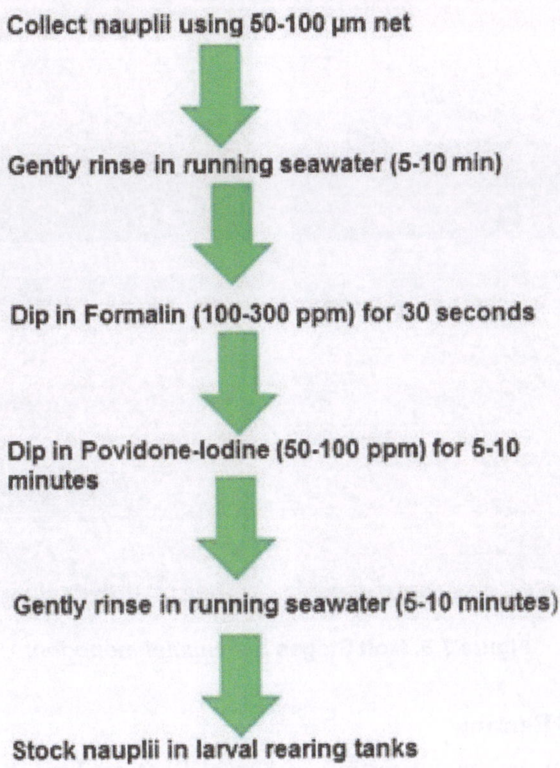

Figure 7.7: Disinfection Procedure for Naupii in Penaeid Hatcheries.

Larval Rearing

☆ To maintain optimum water quality, optimum number of larvae should be stocked and water quality is to be monitored by proper exchange of water.

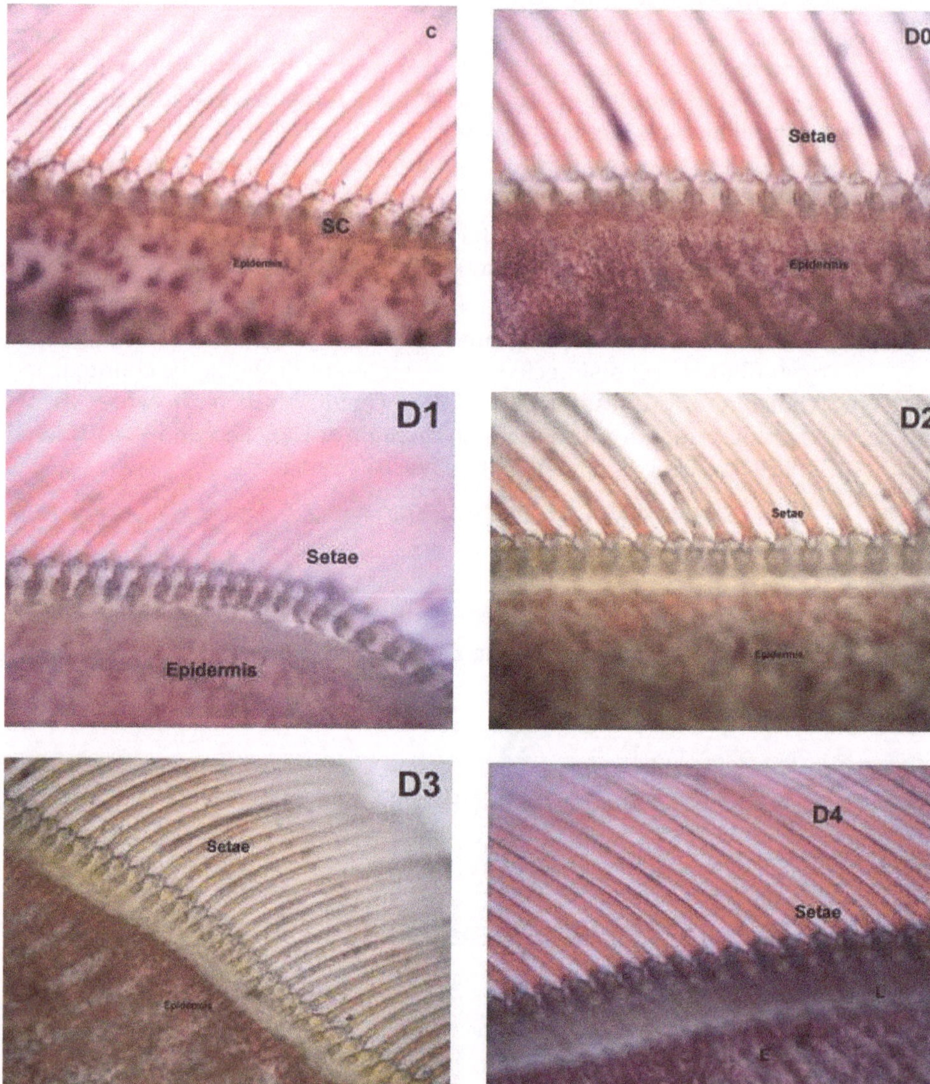

Figure 7.8. Molt Stages in *Penaeus monodon*.

BMP for Larval Rearing

☆ Stocking in the entire unit of Larval Rearing Tank (LRT) should not exceed 3-4 days to maximize biosecurity and prevent management related problems.

- ☆ Water should be exchanged at 10-30 per cent per day through the 4-6 day mysis period, 30-50 per cent per day from PL1-5 and at >50 per cent per day from PL6 until harvest at PL15. If any disease or water quality problems occur, rate of water exchange should be increased.
- ☆ In case of any disease symptoms appears, samples should be sent to a PCR laboratory once (2-3 days before harvest) or twice (at nauplius and PL5 also) during the cycle for screening for viral diseases.
- ☆ Water exchange and algal feeding follow a pattern as the larvae pass through protozoea, mysis and metamorphose in to post larvae and an optimum schedule should be followed.
- ☆ In case of any disease symptoms appears, samples should be sent to a PCR laboratory once (2-3 days before harvest) or twice (at nauplius and PL5 also) during the Water exchange and algal feeding follow a pattern as the larvae pass through protozoea, mysis and metamorphose in to post larvae and an optimum schedule should be followed

Live/Fresh Feed and Feeding Management

Live feed is the crucial component for the successful hatchery production. In penaeid hatcheries live/fresh feed includes microalgae, Artemia and broodstock diets. Optimization of feeding regime is imperative to keep the water quality at optimum levels and production performance

Microalgae

Generally, *Chaetoceros, Skeletonema, Thalassiosira* or *Tetraselmis* spp. are used as live feed and every hatchery maintains an indoor and outdoor facility with qualified technician to carry out the work. The outdoor mass culture is the most vulnerable area for the high level proliferation of bacterial population.

Better Management Practices

- ☆ For zoea, the best feed is clean live algae (*Chaetoceros* sp., *Thalassiosira* sp. or *Skeletonema* sp.), fed at 80-130,000 cells/ml. Alternatively, dried spirulina and liquid or powdered/microencapsulated dry diets (10-80 microns in size) can be fed every 2-4 h if live algae are not available, especially during failure in the outdoor algal unit.
- ☆ The algae in full bloom are concentrated through filtration through a fine mesh bag and the concentrate added to the larval rearing tanks. The final cell densities should be about 80-130,000 cell/ml for zoea and mysis (peaking at Z3), declining to 50-60,000 cells/ml during early Pl stages (optional)
- ☆ The pure culture must be screened from time to time to avoid any contaminations which otherwise creeps in unnoticed and later becomes a problem.
- ☆ Outdoor tank contamination should be avoided by maintaining good water quality, preventing air borne dust and through use of proper nutrients and inoculum

Artemia

Artemia nauplii is one of the most important live feed in the rearing of shrimp larvae from mysis stage onwards and is a major source of contamination. About 6 to 7 kg of artemia cyst is needed for the production of each million tiger shrimp PL. Feeding of artemia should be carried out 4 to 5 times daily, and always 3-5 larvae of artemia should be available in the larval rearing tank. There is a possibility of contamination due to artemia cyst outer covering and artemia disinfection is not carried out by all the hatchery operators. Hatching of artemia is influenced by the salinity, light intensity and aeration. Further when the nauplii are being fed, the hatched outer covering could contaminate the larval rearing tanks.

Better Management Practices

- Decapsulated artemia should be used for hatching.
- Bacterial load in the artemia should be avoided through proper disinfection protocol
- Artemia nauplii, which are active only should be collected by using their positively phototactic behavior

De-capsulation of Artemia Cyst

Artemia has been major vector for several pathogens such as bacteria, fungi *etc.* Therefore decapsualation or removal of the cysts is recommended to increase hatching rate and avoid contamination of Artemia culture water and larval rearing tanks with pathogen and other organic matter that can lead to disease and mortality.

Procedure

- Requirements for decapsulation of 1 kg Artemia
- Decapsulation vessel: 20 L
- Chlorine bleach: (NaHOCl; 5.5 per cent); 8 L at 2-10ºC
- Sodium Hydroxide: (NaOH; 40 per cent); 4 L at 2-10ºC
- Harvest bag: 100 μm (250 mesh)
- Place the required quantity of cysts in plastic containers (1 g cyst per 15 ml)
- Add freshwater and hydrate with continuous aeration (1 h)
- Collect the hydrated cyst into an 80 to 100 μm cyst (250-320 mesh) and wash under running clean fresh water (preferably) or in seawater
- Pour chilled sodium hydroxide into the decapsulation vessel along with hydrated cysts
- Add chilled bleach and provide vigorous aeration until cysts turn bright orange. Orange colour indicates the completion of process; cyst buoyancy also indicates the completion of decapsulation. If 90 per cent cyst sinks, it indicates the completion of process. The time taken for decapsulation varies according to the temperature. Close monitoring is required as over

exposure in the decapsulation solution adversely affect the hatch rate. Standardization is required for particular conditions.

Brood Stock Diet

Broodstock nutrition is the major factor on spawn quality

- ☆ Fresh feed like polychaetes, squids, and mussels are known to be a major source of contamination

Better Management Practices

- ☆ Fresh/frozen SPF polychaetes can be procured to reduce any risk of infection creeping in through live wet feeds
- ☆ In the absence of SPF polycheate, the live feed should not be from polluted/ contaminated source
- ☆ Standard artificial brood stock diets should be used.
- ☆ The live polychates and mollusc should be depurated to reduce the level of contamination before feeding.
- ☆ Screening after depuration should be done from time to time to ensure the quality of these live feed.
- ☆ Should not be fed with brachyuran crabs

Management Measures in Food-Safety Issues

During the early stage of development of penaeid shrimp hatchery technology, use of antibiotics as a prophylactic measure was an integral part. Later when food safety issues cropped up with the presence of antibiotic residues in adult farm reared shrimps, the use of certain antibiotics has been banned in hatcheries. Probiotics is gaining acceptance day by day, however proper method and the optimum dose for the strain of probiotics used are not given importance thus reducing probiotic effect and increasing the cost of production.

Better Management Practice:

- ☆ No antibiotics should be used for prophylactic purposes in larval rearing.
- ☆ Probiotic based seed production should be followed
- ☆ A good probiotic (high cell count (>10^{8-9}cfu/g of multiple bacterial strains) should be used from zoea stage 1 throughout the larval and post-larval rearing stages and the concentration in water should be appropriate (aprox.10^{5-6}cfu/g)

Hatchery Discharge Water Treatment

Aquaculture activities contaminating the source water is one of the issues in all aquaculture ventures. The hatchery operation should ensure that discharge water from hatcheries meet all criteria for discharge water quality.

Better Management Practices

- ✰ All hatcheries should have a three level waste water treatment system for effective treatments.
- ✰ ETP procedure should be strictly followed with waste water being disinfected with proper disinfectant (like bleaching-60 ppm) to ensure that it does not cause any adverse impact on environment.

Shrimp Seed Quality Assessment

Though PCR testing becomes common with all the hatcheries, a holistic approach for total assessment of the shrimp seed quality is essential to ensure good quality seed.

Better Management Practices

The holistic quality assessment is to be done at 3 levels

Level 1

Observation of animal and environment with examination based on gross features is to be performed on a routine basis. It includes health examination of broodstock, sex determination, moult and ovarian development staging, removal of sick/moribund individuals, selection of nauplii by phototactic response, zoea/mysis stage feeding activity by observation of faecal strands, larval activity, postlarval health, activity and behaviour, stress.

Other aspects like good swimming activity, active phototaxis, faecal string, luminescence, stage homogeneity, intestinal contents are also important quality parameters to be judged. A preliminary examination of Post Larvae (PL) in the tank is made to assess size distribution, benthic behaviour, swimming activity, feeding and colour. Then a sample is examined more closely:

Level 2

More detailed examination using light microscopy examination of larval and post larval quality and squash mounts, with or without staining, and basic bacteriology of shrimp, feed and water should be included. For microscopic observation, 20-30 PL are to be randomly selected and analysed the condition of the hepatopancreas, gut content, necrosis, deformities, epibiont fouling should for confirming quality of the PLs.

Level 3

Screening of broodstock, nauplli and postlarvae should be done by molecular techniques and immunodiagnostics (*e.g.* PCR). For *P. monodon* post larvae, WSSV is checked whereas for the *L.vannamei* WSSV, IHHNV along with other OIE listed virus should be screened. In addition, vibrio examination should be done to check for potentially harmful *Vibrio* sp. bacteria in the PL. For this purpose, a random sample of 100 PL are taken from the tank, sterilized externally by dipping in 70 per cent alcohol, washed, ground and then streaked with sterile wire loops, or 0.1 ml pipette. The PL pass if the number of green colonies is <60/plate and the number of yellow colonies is <80/plate on the TCBS plates, and if no luminescence is noted in the TSA media.

Reducing Stress during Harvesting and Transportation

The PL should be acclimated in the hatchery to the salinity expected in the on-growing farms. This is to reduce the stress on stocking in the ponds, a critical point in the process which can lead to high mortalities.

General Test

Good quality post larvae should be selected through observation and microscopic, molecular and stress testing. The 6 criteria used are scored using a standard score and are: (i) hepatopancreas (ii) gut condition (iii) fouling (iv) Deformity (v) muscle : gut ratio (vi) MBV. For the stress test the PL must be at least 10 mm long (>PL8-10) withstand these tests, and it is ideal to test it as close to stocking date as possible – *i.e.* PL 12-15 at >12mm total length.

Stress Test

Stress testing is done using two stress factors of salinity and formalin.

Survival (per cent) = (No. active PL/total PL in container)* 100

Scoring: the PL batch passes if the survival >75 per cent and fails if below score.

PCR Testing

PCR Testing for WSSV can help reduce the risk of crop loss due to this disease and should therefore be checked on each batch of PL stocked.

PL Harvest and Transportation

Transportation should be stress free and density should be maintained as per the duration of transportation. Salinity adjustment can be done by adding freshwater to gradually achieve a salinity change of <3ppt/hour from 30 -20 ppt, but should be reduced to <1ppt/hour from 20-10 ppt and <0.5ppt/hour from 10-5 ppt. Such salinity adjustments should commence only when the PL are older than PL 10 (preferably 2-3 days before harvest), when their gills are fully developed (looks like Christmas tree) and they are able to tolerate such rapid salinity changes.

Conclusion

Best Management Practices (BMPs) in the hatcheries are essential for producing qualityseed. Unless quality seed is ensured by adopting BMPs, the whole production-consumption system in the value chain will be affected. Though many attempts have been made to develop and disseminate BMPs, due to location specific problems, these have not been widely implemented. There is a general lack of qualified staff in the hatcheries and awareness on the technical details for implementing these BMPs. Ultimately, what we require is a twin approach of developing a manual with specific details for all the 15 BMPs identified in this publication and a framework for analysing location specific problems that would require additional measures to ensure that these BMPs are implemented. The present publication is aimed at highlighting such location specific issues that need to be examined and drawing attention to specific bottlenecks in producing quality seed.

Scientific Grow Out Shrimp Culture

Shrimp farming in India is dominated by exotic white shrimp, *P.vannamei*, Indian white shrimp *P.indicus* and black tiger shrimp, *P.monodon*. Penaeid shrimps were part of traditional farming systems based on tide fed system started in early 1970s. The productivity of this tide fed systems were very low and along with brackishwater finfishes and penaeid shrimps contributed a productivity of 500 kg per ha. However, after the successful penaeid shrimp seed production trial and massive development of shrimp hatchery, availability of shrimp seed doesn't became a scarcity with the development and refinement of shrimp farming techniques, scientific farming started in late 1980s with a productivity above 2 -5 ton per ha. Viral disease outbreak and environmental issues remain always a challenge to successful shrimp farming. In this context, it became utmost important to follow scientific better management practices to ensure sustainable crop to the farmers. Success of any shrimp culture depends on better management practices involved in pond preparation and pre-stocking management steps. Pond preparation is one of the most important pre stocking management measures essential for optimum growth of shrimp in grow out farming systems. There are various points to be taken care during the pond preparation for shrimp culture.

Drying the Pond Bottom

After each harvesting cycle, the pond bottom is allowed to dry and crack. It helps to oxidize the decomposed organic components, leftover in the pond after the previous culture. Generally pond bottom is allowed to dry for 7-10 days and, it allows soil crack to a depth of 25-50 mm. It helps to reduce the risk of disease outbreaks and improve shrimp production.

Ploughing or Taking

Ploughing or raking the pond bottom help to exposes the nutrient rich sub soil and fast mineralization and oxidation of the organic compounds and harmful gases. Tiling and ploughing is not generally recommended in acidic soils as it increases the soil pH.

Top Soil Removal

The top black soil and bottom sludge to be removed to prevent development of anaerobic condition during culture period. The sludge must be disposed away from the pond site, so that it does not seep back into ponds. Grow out pond with high stocking density, entire pond top soil is removed where as modified extensive ponds, areas of the pond where there is a high accumulation of organic matter from previous crops, such as feeding zone should be removed.

Undrainable Ponds

Undrainable or ponds with acid sulphate soil, sun exposure can cause iron pyrite get oxidized into its suphide forms and subsequent reduction in soil pH drastically. These types of soil can be prepared after repeated flushing of water intake and out followed by soil lime treatment to increase soil pH.

Top soil removal

Pond bottom after top soil removal

Figure 7.9. Pond Bottom Top Soil Removal.

Liming

During pond preparation liming is applied to optimize pH and alkalinity conditions of soil and water. The type and amount of lime to be added depends mainly on the soil and water pH, which should be checked before lime application. The recommended levels of lime application during pond preparation are given in Table 7.1. The soil and water pH can be measured with a pH meter. Generally agricultural lime or dolomite can be applied if soils of pH >5andQuick lime or hydrated lime can be applied if soil pH below 5. Where disinfectants like bleaching powder (calcium hypochlorite) is used, applies lime only 3-4 days after the application of disinfectant as lime reduce the effectiveness of the disinfectant.

Table 7.1. Amount of Lime (tons/ha) to Raise the Soil pH to 7.0

Soil pH	*Quantity of Lime Material (tons/ha)*		
	Dolomite	*Agricultural*	*Quick Lime*
6 to 6.5	5.7 to 2.8	5.5 to 2.8	4.6 to 2.3
5.5 to 6.0	8.5 to 5.7	8.3 to 5.5	6.9 to 4.6
5.0 to 5.5	11.3 to 8.5	11.1 to 8.3	9.2 to 6.9
4.5 to 5.0	14.2 to 11.3	13.9 to 11.1	11.5 to 9.2
4.0 to 4.5	17.0 to 14.2	16.6 to 13.9	13.8 to 11.5

Water Intake

Stringent measures to be followed to prevent entry and growth of any unwanted and pathogenic agents in culture ponds. It can be achieved via proper filtration of intake water using appropriate mesh screens, disinfection of intake water. Filter water first through coarse screens to remove larger aquatic animals and debris and then pump into a supply/settling canal for allowing the suspended particles in the water to settle. Then, pass the water through a series of progressively finer

Lime applied on pond bottom

Lime application in pond water

Figure 7.10. Lime Application in Pond.

screens (150– 250 µm mesh size) before introducing into the reservoir. Chlorinate water in the reservoir with sufficient chlorine (10 ppm) to kill any potential vectors or carriers in the source water. For one ha reservoir pond of one meter depth, 150-160 kg of calcium hypochlorite providing 65 per cent active chlorine would give a final concentration of 10 ppm. However, it is advised to calculate the actual dose based on the chlorine demand of water and actual chlorine content in bleaching powder. Vigorously aerate reservoir at least 48 h for de-chlorination to remove residual chlorine.

Optimum water quality criteria for intake water are given in Table 7.2. Keeping a suitable reservoir also facilitate chemical treatment to reduce disease outbreak and to make water management more effective during production cycle. The water parameters that should be monitored routinely in ponds during culture period are temperature, pH, salinity, dissolved oxygen and transparency.

Table 7.2. Optimum Water Quality Criteria for Intake Water

Sl.No.	*Parameter*	*Normal Range*
1.	Temperature (°C)	25-33
2.	Salinity (ppt)	10-34
3.	pH	7-9
4.	Transparency (cm)	25-50
5.	Dissolved Oxygen (ppm)	4-6
6.	Total Alkalinity (ppm)	50-300
7.	Nitrate- N (ppm)	< 0.03
8.	Nitrite- N (ppm)	< 0.01
9.	Ammonia- N (ppm)	< 0.01
10.	Heavy Metals (ppm)	Nil to 0.0001

- ✰ The pH should be in optimum level of 7.5 to 8.5 and should not vary more than 0.5 in a day.
- ✰ Variations in salinity not exceeding 5 ppt in a day will help in reducing stress on the shrimp.
- ✰ The optimum range of transparency measured using secchi-disc is 25-35 cm. Total Ammonia Nitrogen (TAN) and nitrite N concentration should not be more than 1 and 0.5 ppm, respectively.
- ✰ Any detectable concentration of hydrogen sulphide is considered undesirable. Periodical exchange of chlorine treated water from reservoir as and when required will help in maintaining the water quality in optimal range.
- ✰ The use of aerators results in mixing of water at surface and bottom and breakdowns the DO and thermal stratification.
- ✰ Use of inputs without proven efficiency should be strictly avoided. The discharge water from the shrimp ponds has to be allowed into a treatment system pond before letting it into the environment so that the suspended solids may settle at the bottom.

Soil Quality Parameters

The soils with moderately heavy texture (sandy clay, sandy clay loam and clay loam), electrical conductivity value of 4 dS m-1 or more, pH ranging between 6.5 and 7.5, organic carbon content of 1.5 to 2 per cent and calcium carbonate content of more than 5 per cent are the best suited for shrimp aquaculture. In order to understand the condition of the pond bottom, soil pH, organic matter and redox potential (Eh) for oxidized/reduced pond bottom condition have to be monitored regularly. The Eh of pond sediment should not exceed -200 Mv. Redox potential also called as oxidation-reduction potential (ORP) is an important parameter to measure the relative degree of oxidation and reduction in aquaculture ponds. ORP in the pond soils decreases towards reduction (more –ve side) with progress of culture period. Though highly variable, ORP values denoted by Eh and quantified in milli volts (mV) are best used as an indicator to understand relative status of the soil.

Fertilization of Pond Water

The purpose of fertilization is to ensure the growth of primary producers in culture ponds. They initiate natural food web in the aquatic ecosystem and directly or indirectly contribute shrimp growth also. It also helps to maintain desirable level of transparency (25-40 cm) which prevents development of harmful benthic algae. Phytoplankton in culture ponds also help to improve the water quality parameters in grow out ponds.

Fertilizers can be applied depending on the fertility status of the soil. Organic fertilizer like dry cow dung at the rate of 500 – 2000 kg ha^{-1} and inorganic fertilizers like urea and single superphosphate (SSP) at 25 – 100 kg ha-1 can be applied depending on the organic carbon content (1.5-2.5 per cent), available N (50-75 mg/100 g soil) and available P (4-6 mg/100 g soil) content in the pond. Of the original dose, 10 per cent can be applied fortnightly to maintain the desired level of algal bloom. The Secchi disc transparency should be in the range of 25-40 cm. Brownish green colour of pond water indicate that the pond is ready for stocking the seed.

Application of Organic Juice

Yeast based organic preparation are recently being used in zero water exchange shrimp culture ponds as a probiotic. It can be prepared using ingredients like 30 kg paddy flour, 30 kg molasses and 4 kg yeast *Saccharomyces cerevisiae*. Allow these ingredients to get ferment in 48 h and can be applied in one ha pond. Organic juice can be applied in biweekly interval to improve the fertility status of water @ 1-2 ppm.

Figure 7.11. Shrimp Pond Water.

Table 7.3. Commonly Used Nitrogenous and Phosphatic Fertilizers in Brackishwater Shrimp Aquaculture

Fertiliser	*Availability*
Ammonium sulfate $(NH_4)_2SO_4$	20-21 per cent as NH_3
Ammonium nitrate (NH_4NO_3)	17-18 per cent as $NH_3$17-18 per cent as $N0_3$
Urea (NH_2CONH_2)	46 per cent N
Single super phosphate [Ca $(H_2PO_4)_2$	16-18 per cent P
Ammonium phosphate $(NH_4H_2PO_4)$	48-56 per cent P

Seed Selection and Stocking

Stocking shrimp is perhaps one of the most important components of a bio-security program. Use seeds produced from domesticated shrimp stocks that are free of specific diseases ("Specific Pathogen Free" or SPF) and or with stocks resistant to specific disease agents (SPR) SPF broodstock from certified Nucleus Breeding Center (NBC). These are bio-security facilities where there are two or more years of documented disease testing to support their SPF status. Before purchasing, shrimp post larvae should be checked for their general condition such as activity, color, size, *etc.* If there is any dead and abnormal colored PL in the stock, the entire batch should be rejected. Before stocking at the pond, PL should be treated with formalin at 100 ppm concentration for 30 minutes in well aerated tanks to remove weak PL. Maintain a balanced or optimum stocking density is also an important component of bio-security programme.

Pond Bottom and Water Quality Management

Disease outbreaks in shrimp grow out culture is directly related to pond bottom and water quality. Zero water exchange must be followed throughout the culture period to prevent the disease occurrence. Use chemical like $KmNO_4$ dip treatment may be followed to disinfect all equipment– screen net, cast net, trays or accessories while sampling and regular monitoring (surveillance) of shrimp stocks in bio-secure culture facilities is a necessary component of a bios-ecurity plan.

Mineral Requirement

In addition to basic water quality parameters, the mineral profile of water is important in *P.vannamei* culture. Actual mineral requirements are difficult to quantify due to the variability in the ionic profiles of pond waters. The bioavailability of minerals will be a function of their concentration in water. In general, levels of minerals in pond water have to be more or less similar to the levels in seawater diluted to the same salinity. However, pond water characteristics of even the very closely located ponds will not be exactly the same. Ionic ratios are quite different between seawater and different source waters. The ratio of Na (Sodium) to K (Potassium) and Mg(Magnesium) to Ca (Calcium) in water appears to be more important than pond water salinity. Improper ratios of these minerals in water lead to osmotic stress which has a cascading effect on growth and survival of shrimp. The Na:K and Mg:Ca ratios should be preferably 28:1 and 3.4:1 respectively. It should be noted that although high levels of Ca also seem to be necessary, the ratio of Ca:K, which is about 1:1 in seawater, may also be important. In waters where the Ca:K and Na:K ratios are high, the addition of K to reduce these ratios in low saline waters increases the shrimp growth.

A number of commercial products are available in the market(potassium chloride, magnesium chloride, mineral mixtures *etc.*) to correct ionic ratio imbalances in pond water. Many commercial products do not mention the composition of minerals. Utmost caution should be exercised while using such commercial products since studies have indicated lack of standard and efficacy. Dose (g/m^3) of a product required to be applied to the pond for a particular mineral may be calculated using the following formula.

Dose (g/m^3) = Desired concentration of particular mineral (mg/L) ÷ Percentage of that particular mineral in salt/100. For example, if you want to use magnesium sulphate (Epsum salt) containing 10 per cent Mg to increase Mg concentration by 25 mg/L: Dose of salt = 25 ÷ 10 per cent/100 = 250 mg/L. Generally under high or low saline waters with optimum mineral concentration and proper ionic ratios, there is no need of supplementation. However, throughout the culture period, major minerals are lost due to soil adsorption, shrimp harvest, draining at harvest and seepage, altering their concentration. Hence, there is a need to evaluate the mineral concentrations in pond waters regularly and to supplement them in case of any deficiency.

Better-Feed Management

Cost of feed accounts for about 40 per cent to 60 per cent of the total production cost. Do not use fresh feed, trash fish, bivalves *etc.* as it can carry vectors. Feed monitoring should be done with check tray evaluation for optimum feed management. Feeding area can be shifted at least once in 7 to 10 days depending on the bottom condition along feeding area. Reduce feeding during periods of low

DO, plankton crash, rain fall, extremes of temperature *etc.* Slightly under feeding is better than over feeding, which saves money and reduce disease risks and during disease outbreaks. Proper storage of feed is also an important component in the bio-security plan.

Shrimp Health Monitoring

Shrimps should be sampled once in a week by cast netting and should be checked for their general health conditions like external appearance. For example, a pale whitish gut showed gut infection while a normal gut will have a light or golden brown colour. Probiotics, immune stimulants, bio-remediating agents can be employed as prophylactic measures in grow out culture. Yeast based organic preparation (60 kg rice flour, 30 kg yeast and 3 kg yeast) application can be applied to improve the overall pond microbial balance. Since there is a serious concern on the use of antibiotics, their use in shrimp farming should be avoided. Avoid unnecessary handling (touching). Control workers' movement in and across the farm and minimize number of workers in stocking, harvest, sampling *etc.* It is utmost important to make environmental cleanliness and control Human traffics – guest, workers, technicians and movement across the farms. Educate people on bio-security.

In intensive shrimp culture ponds, the accumulated black sludge in the pond bottom leads to reduced conditions and negative redox values. CIBA studies have revealed that ORP value of -200 mV or more is not desirable during the culture period. This problem can be avoided by providing proper aeration uniformly throughout the pond by placing aerators at right place, chain dragging along the pond, central drainage system and disposal of sludge through heavy duty pumps.

Farm Record Maintenance

Records are necessary to identify various risks and to rectify these problems at the earliest during the production cycle. Record keeping also helps the farmer to learn from past mistakes, thus reducing risk and costs of production in subsequent crops.

Better Practices for Harvesting

Harvesting must be avoided during moulting period and agricultural lime can be applied 3-4 days before harvesting. Try to do harvesting in the early morning or evening. Harvesting should be done with dragnet with minimum delay. After harvesting transport the crates with crushed ice at 1:1 ratio for better preservation. Nutrient rich pond effluent must be treated before get discharged into water source.

References

1. Allan, G. and Burnell, G. eds., 2013. Advances in aquaculture hatchery technology. Elsevier.
2. Balasubramanian, C. P., ShyneAnand, Kannappan S and Biju I. F. Training manual onrecent advances in farming of pacific white shrimp (Penaeus vannamei), Training manualseries, no-14, 2018, 126 pp

3. Dall, W.H.B.J., Hill, J., Rothlisberg, P.C., Sharples, D.J., Blaxter, J.H. and Southward, A.J., 1990. Advances in marine biology. Academic Press.

4. FAO (Food and Agriculture Organization of the United Nations) (2007) Improving Penaeus Monodon Hatchery Prac-tices. Manual Based on Experience in India. FAO Fisheries Technical Papers T446. FAO, Rome

5. Ma, K.Y., Chan, T.Y. and Chu, K.H., 2011. Refuting the six genus classification of Penaeus sl (Dendrobranchiata, Penaeidae): a combined analysis of mitochondrial and nuclear genes. Zoologica Scripta, 40(5), pp.498-508.

6. Motoh, H., 1981. Studies on the fisheries biology of the giant tiger prawn, Penaeus monodon in the Philippines. Aquaculture Department, Southeast Asian Fisheries Development Center.

7. Mohamed, K.H., Muthu, M.S., Pillai, N.N., Ahamed Ali, S. and Pandian, S.K., 1983. A simplified hatchery technique for mass production of penaeid prawn seed using formula feed. Indian Journal of Fisheries, 30(2), pp.320-332.

8. Muthu, M.S. and Laxminarayana, A., 1982. Induced maturation of penaeid prawns-A review. Proceedings of Symposium on Coastal Aquaculture. Part 1, Prawn Culture. Marin. Bioogical Association of India, pp. 16-27

9. Parado-Estepa FD, Quinitio ET, Borlongan E. 1991. Prawn hatchery operations. AquacultureExtension Manual No. 19, SEAFDEC Aquaculture Department, Iloilo

10. Saraswathy, R., Kumararaja, P., Lalitha, N., Avunje, S. and Muralidhar, M., 2018. Training Manual On Recent Advances in Soil and Water Management in Brackishwater Aquaculture

11. Silas, E.G., Mohamed, K.H., Muthu, M.S., Pillai, N.N., Laxminarayana, A., Pandian, S.K., Thirunavukkarasu, A.R. and Ahamed Ali, S., 1985. Hatchery production of penaeid prawn seed: Penaeus indicus. CMFRI Special Publication, 23, pp.1-41

12. Treece, G.D. and Fox, J.M., 1999. Design, operation and training manual for an intensive culture and shrimp hatchery. DIANE Publishing.

13. Wyban, J., 2013. Biosecurity measures in specific pathogen free (SPF) shrimp hatcheries. In Advances in Aquaculture Hatchery Technology (pp. 329-338). Woodhead Publishing.

Chapter 8

Development of Production System for Mud Worm Polychaete, *Marphysa gravelyi*

S. Kannappan, S. Sivagnanam, C.P. Balasubramanian and K.K. Vijayan

Crustacean Culture Division, ICAR-Central Institute of Brackishwater Aquaculture, 75, Santhome High Road,MRC Nagar, Chennai, Tamil Nadu-600 028

Polychaete worms are aquatic annelids under the diverse group of Phylum Annelida (Read *et al.*, 2019) living in all the environments from the world's oceans to shallow estuaries and rocky shores. As per the *"World Register of Marine Species"*, it has been reported that 11,456 species of polychaetes (1417 genera, 85 families) exist since 1758. It is predicted that the total number of polychaete species in the world would be around 16,700 numbers in this century (Pamungkas *et al.*, 2019). There are about 400 species have been reported from India (Glasby *et al.*, 2000). Polychaetes shows a chiefrole in the ecosystem as feed for marine fish larvae since they endure low salinity and oxygen levels and able to adapt in sediments containing heavy metals associated with their food. They usually detoxify heavy metals as less hazardous in the gut by absorption through the body wall. Few species of polychaetes can live in sediments with high trace metal concentration and their body do not replicate sediment due to regulation by polychaete worms (Dean, 2008).

Shrimp and fish hatcheries have been extensively using polychaete worms as feed for brooders for their effective maturation and production of high-quality juveniles as they have high content of polyunsaturated fatty acids (PUFA) especially arachidonic acid, eicosapentaenoic acid and docosahexaenoic acids (Leelatanawit *et al.*, 2014). Polychaete worms about 60 species are being used as bait (the most popular families are Arenicolidae, Eunicidae, Nereididae and Onuphidae) (Cole *et al.*, 2018). Persistent collection of polychaete worms from natural resources are used as feed for fish or shrimp and also as baits for birds would deplete their population (Olive, 1993). Further, it is also opined that depletion of polychaetes in their habtitat diminishes fish and other marine population. Excavation of soil for collection of Polychaete worms disrupts the feeding activity of shore birds due to reduction in availability their prey. Collection of polychaete worms perceived as a non-sustainable activity with potentially detrimental effects on the natural environment. The natural supply of polychaete worms has not been sufficient to meet market demand. Hence, a technology is to be developed for culturing the macrobenthic polychaete worms for rendering food for mariculture practies.

Collection of Polychaetes

The mud samples are to be collected from the various locations of lagoon using a hand corer (Figure 8.1). The polychaetes should be separated from the collected mud samples then counted, and identified using the method described by Fauchald (1977). (Biota: Animalia, Phylum: Annelida, Class: Polychaeta, Sub class : Errantia, Order: Eunicida, Family: eunicidae, Genus: *Marphysa*, Species: *gravelyi*). The physico chemical parameters of the water from the collection area are to be recorded by standard protocols. The worms are transported to the laboratory with a moist substrate for observing morphometric characters and then stocked in the respective tanks.

Preparation of Sand Bed for Polychaete Culture

Fibre-glass reinforced plastic (FRP) tanks having size of 100 cm x 50 cm are placed with plastic couplers in 4-5 Nos. at 10 cm height from the bottom. Aluminum substratum at 10 cm height is to be placed on the plastic couplers (Figure 8.2). Culture bags made up of nylon cloth with 50 μ having the size of 80 cm x 50 cm are placed on the substratum. Sea sand and lagoon based clay soil are collected and sieved using 500-250 μ and 125- 63 μ mesh respectively. Both the sand and clay are mixed at 60:40 ratio and cleaned with potable freshwater and then by seawater (10 PSU,Practical salinity unit). After cleaning, the soil is to be sun-dried for 3 days. Later the mixture of sandbed is collected in the culture bags and arranged in the FRP tank for 10 cm height. Air-lifting sea water recirculation system is to be provided (Poltana *et al.*, 2007) through PVC square frame.

Development of Larval Stages

Gelatinous egg masses of *M.gravelyi* should be collected with the substrate in separate containers from the lagoon for larval rearing. The developmental stages are to be observed, and no feed is given as they are lecithotrophic in nature.

Figure 8.1. Collection of Polychaete Worms from the Lagoon.

Culture of *Marphysa gravelyi*

Live adult polychaete worms of both the sexes in the ratio of 1:1 with the size of 0.8-2.0 g in weight and 10-15 cm in length are collected and stocked in FRP tanks. They are being cultured for 120 days. They can be fed with any commercial shrimp feed after soaking overnight and passing through 50 μ filter. The prepared feed are sprayed twice a day at the rate of 1 per cent of their total weight. Water exchange is to be carried out in 15 days interval. The polychaetes are cultured for 4 months. Various water quality parameters such as salinity 25-30 PSU, pH 7.8–8.2 water temperature 21-27°C are maintained.

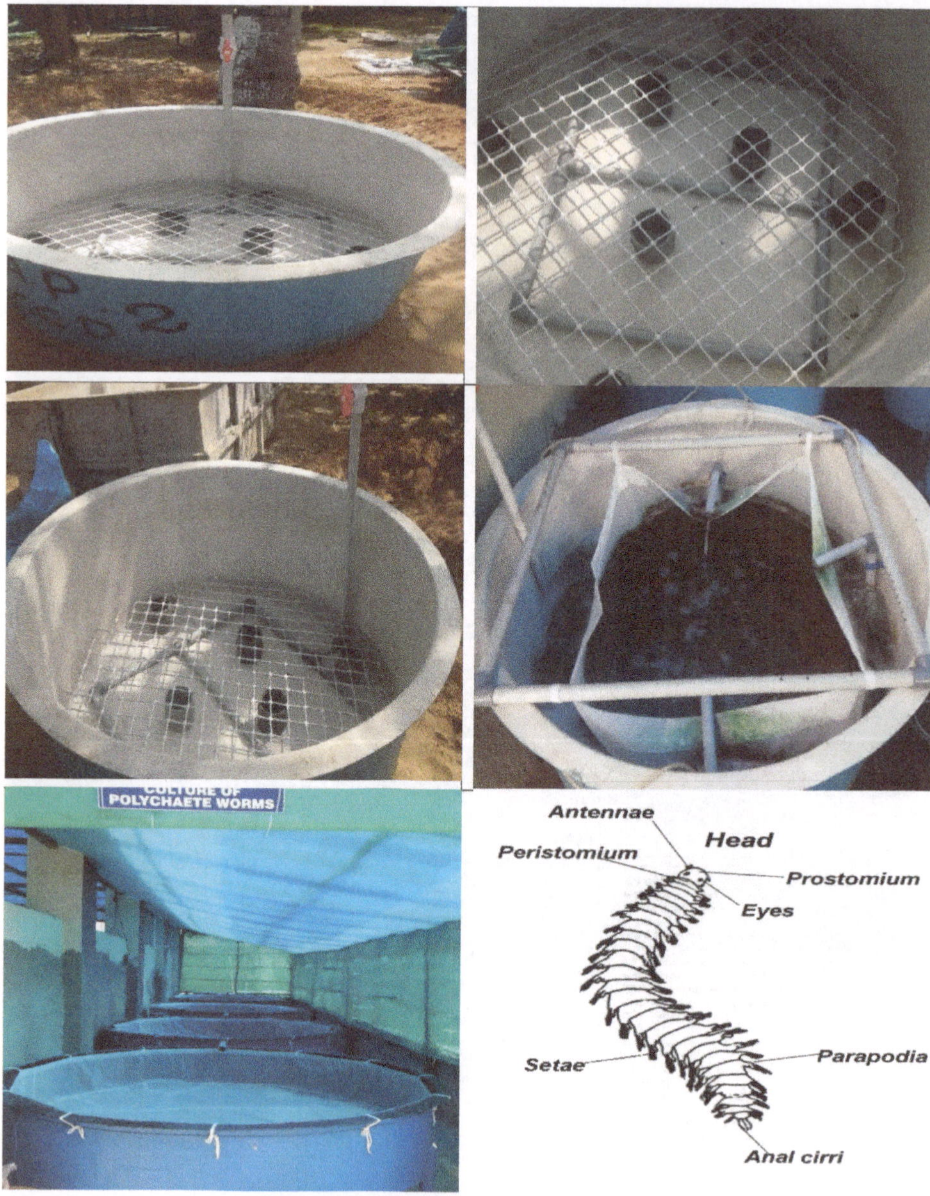

Figure 8.2. Development of Tanks with the Substrate for Grow-Out Practices of Mud Worm Polychaete, *Marphysa gravelyi.*

Plastic tubs having capacity of 25 L filled with sea sand and clay soil at the ratio of 60:40 up to 16 cm height can also be used as sand bed for culture of polychaetes. The water level is to be maintained to 10 cm from the top of the sand bed with aeration. Few polychaetes, say 5 numbers (2 male and 3 female) of young individual worms with the size of 6-7 cm size can be stocked for culture.

Mass Culture of *M. gravely*

Preparation of Sand Bed

FRP tanks with the capacity of 1000 L are arranged with 15 numbers of couplers at 5.0 cm height from the bottom. On the couplers, bamboo grill plate is to be placed. Above this, 2 m nylon cloth bag with mesh size of 100 μ is tied. Sand bed are being prepared using 60 per cent collected river sand sieved through 75 μ filter and retained by 250 μ filter and 30 per cent of sieved clay soll using 100 μ filter and retained by 63 μ filter. Before preparing the bed, first both the soils should be treated with 10 ppm sodium hypochlorite solution for three days followed by washing in freshwater and sun-dried. Water is filled to a height of 10 cm from bottom to the soil bed. Air-lifting flow-through system in three sides of sand bed are to be provided. Five hundred worms with a size range of 9 to 10 cm of both the sexes in the ratio of 200 males and 300 females are stocked. The worms will be fed everyday with 3 ml *Skeletonema* costatum cells (10^6 cells/ml @ 30 per cent body weight) (Bharathidasan *et al.*, 2017) and cultured for four months.

Collection of *M. gravelyi*

Polychaete worms are collected and the species *Marphysa gravelyi* under the family Eunicidae is to be identified as per the keys (Sunder Raj and Sanjeeva Raj 1984). The salinity, pH and temperature of water from the sampled area are to be recoreded for maintaiing the same range of parameters in the culture system. The preferred size of young worms are 6 – 7 cm while the adults are 10 – 15 cm for culture.

Developmental Stages of *M. gravelyi*

Based on microscopic observations, there are four different developemental stages of larval development can be noticed from a single egg mass (Figure 8.3). The eggs are spherical in shape and fertilized eggs hatch out after 96 h and the larva is known as *Prototrochophore*. After 12 hrs, the larvae reach to the *initial metatrochophore* stage followed by *late metatrochophore* after 96 hrs. *Nectochaetae* stage will be formed after 48 h. Later, the jelly mass disintegrates and larva settles to the sediment. The entire larval development takes place in 8–10 days.

Culture of Larvae and Adult *M. gravelyi*

Individual *M.gravelyi* reaches a size of 12 – 20 cm from the size of 6 – 7 cm after culturing them four months. The worms make extensive burrows in the soil and produces 50-100 egg mass. After two weeks, the larvae reaches 5 - 6 cm length and later grow as adult. In our experiment, we could get a total of 2,500 numbers of *M. gravelyi* worms with the length of 6-7 cm from each tank with 90 per cent survival in 4 months from the initial stocking of forty numbers of live adult polychaete worms of both the sexes in the ratio of 10 males and 30 females having size of 0.8-2.0 g in weight and 10-15 cm in length.

Mass Culture of *M. gravelyi* Larvae

It is carried out by stocking 500 numbers of larvae with size of 6-7cm in 1000 litre capacity FRP tanks with a water re-circulation facility (Figures 8.4 and 8.5).

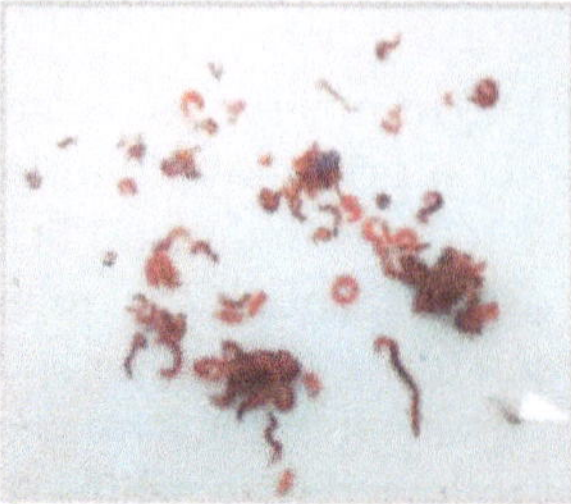

5 - 50 mm
Larvae
1 - 2 weeks

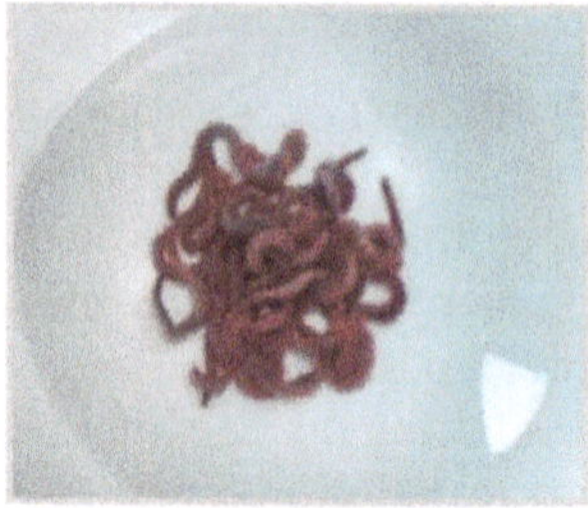

50 - 70mm
Young worms
2 - 3 weeks

100 - 250mm
Adult worms
5 - 6 months

Culturing Polychaete worms
With young worms

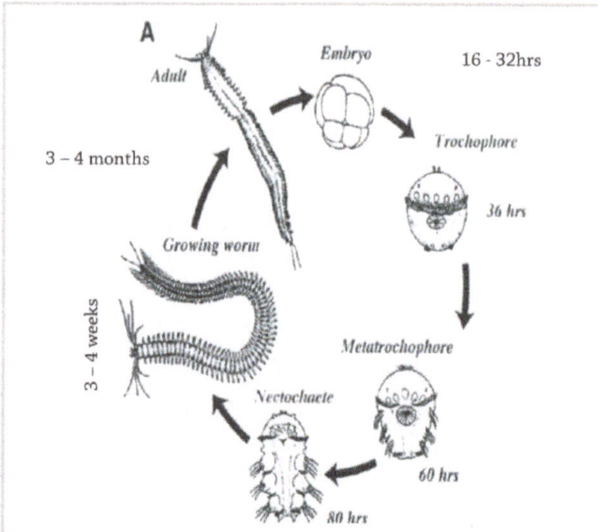

CENTRAL INSTITUTE OF BRACKISHWATER AQUACULTURE
(Indian Council of Agricultural Research)
75, Santhome High Road, Chennai - 600 028.
Tel : +91-44-2461 0565 / 2461 6948 / 2461 8817 Fax : 91-44-2461 0311
E-mail : director@ciba.res.in Website : www.ciba.res.in

Figure 8.3. Culture, Reproduction and Life Cycle of Polychaete Worms, *Marphysa gravelyi*.

Figure 8.4. Grow-out Practices of Mud Polychaete Worm, *Marphysa gravelyi* in FRP Tanks.

The larvae are fed with *Skeletonema* costatum @1 per cent body weight. Most of the larvae will became adult in 4 months (15-17 cm) with a biomass of 3200 Nos. as adult worms from each tank with 80 per cent survival.

Conclusion

The eggs of *M.gravelyi* are contained inside a jelly mass, wherein their larval development takes place. *M. gravelyi* produces polylecithal eggs and hence the larvae are lecithotropic. The entire larval development can be observed in 8-10 days. When

Figure 8.5. Grow-out Practices of Mud Polychaete Worm, *Marphysa gravelyi* in FRP Tanks.

adult *M. gravelyi* worms grown in FRP tanks, 2,500 numbers of young *M. Gravelyi* worms (6-7cm) are obtained as biomass from each tank with 90 per cent survival in 4 months. While stocking *M.gravelyi* larvae (6-7 cm) in 500 numbers in 1000 l FRP tanks with a water re-circulation facility, most of the larvae become adult in 4 months (15-17 cm). A biomass of 3200 Nos. can be harvested as adult worms from each tank with 80 per cent survival.

Since, *M. gravelyi* contains essential nutritional constituents that may be used as a brood stock feed to facilitate reproductive development of aquatic organisms. Hence, grow-out practices of polychaete worms for commercial use are obviously a preferable alternative method against collecting from the wild sources. This species can be mass cultured under a controlled condition and supplied as best live feed for brood fish/shrimps. Mass culture of mud worm, *M. gravely* can prevent its stock depletion from natural populations inhabiting from the coastal lagoon/ backwater/creek *etc.*

References

1. Analytical chemistry of food, 178 pp. Blackie Academic and Professional Press, London and Biochemistry 37: 911-917.
2. Bharathidasan, V., Partha Sarathy, P., Selvaraj, P., and Murugesan, P. (2017). Sexual reproduction and larval development of *Nereididae*, Polychaete, *Perinereis cultrifera* (Grube, 1840) *International J. Sci. Inventions Today*. 6, (6) 772-782.
3. Cole VJ, Chick RC, Hutchings PA (2018) A review of global fisheries for polychaete worms as a resource for recreational fishers: Diversity, sustainability and research needs. Rev Fish Biol Fisher. 28: 543-565.
4. Dean HK (2008) The use of polychaetes (Annelida) as indicator species of marine pollution: a review Rev. Biol. Trop. 56 (4) 11-38.
5. Fauchald,K (1977) The polychaete worms definitions and keys to the orders, families and genera Natural history Museum of Los Angeles County in conjunction with the Allan Hancock Foundation University of Southern California, Science Series 28, February 3.
6. Glasby, C.J., Hutchings, P.A., Fauchald, K., Paxton, H., Rouse, G.W., Watson Russel, C., Wilson, R.S. (2000).Class : Polychaeta. In: Beesley PL, Ross GJB, Glasby CJ (eds) Polychaetes and allies. Thesouthers synthesis. Fauna of Australia, Vol 4a:*Polychaete,Myzostomida,Pogonophora, Echiura,* CSIRO Publishing, Melbourne, PP 1- 296.
7. Olive, PJW (1993) Management of the exploitation of the Lugworm *Arenicolamarina* and the Ragworm *Nereis virens*(Polychaeta) in conservationareas. Aquatic Conservation: Marine and Freshwater Ecosystems, 3 (1): 1–24.
8. Poltana, P., Lerkitkul, T., Pongtippatee, P., Taweepreda, O., Asuvapongpattana, S., Wongprasert, K., Sriurairatan, S., Chavadeji, J., Sobhon, P., Peter, J.W.O., and Withyachumnaankul, B. (2007). Culture and development of the polychaete*Perinereis*cf. *nuntia.Invertebrate reproduction* and *development*. 50, (1) 13 – 20.
9. Read, G., Fauchald, K. (Ed.) (2019). World Polychaeta database, accessed at http://www.marinespecies.org/polychaeta on 2019-08-25.
10. Sunder Raj SK and Sanjeeva Raj PK (1984) Polycaeheta of the Pulicate lake (Tamil Nadu). Journal of Bombay Natural Histroy society.84-98.

Chapter 9

Formulation and Preparation of Fish and Shrimp Feed

J. Ambasankar, Syamadayal and K.P. Kumaraguru Vasagam

ICAR-Central Institute of Brackishwater Aquaculture,
75, Santhome Highroad, R.A. Puram, Chennai–28

Feed cost is the major input cost in aquaculture which assumes 50-70 per cent of cost of production. Hence the feed should be nutritionally well balanced to maximize the growth rate and feed conversion efficiency. Feed formulation is a scientific method of combining available feed ingredients to meet the pre-established nutrient requirements to prepare a compounded feeds for the candidate species intended for aquaculture in a cost effective manner. It is an essential application in nutritional science to a practical situation in feed manufacturing process. The main objective of feed formulation is to utilize the knowledge of nutrient requirements, cost effective available feed ingredients, and their inclusion levels based on their nutritive composition, digestibility and anti-nutritional factors. Feed formulation usually follows a certain sequence of 'trial and error' steps, involving lot of educative guesses and skills. The nutritionist who attempts to formulate a feed must have in depth knowledge of the biology of the target species, feeding behavior and the purpose for which feed is being prepared (De Silva.1989).

Formulation and preparation of compounded feeds is based on several scientific and economic factors.

A. Information on Candidate Species

- ☆ Nutrient requirements of the candidate species including protein, energy, fat, amino acids, fatty acids, vitamins, minerals for each stage of culture.
- ☆ Feeding habits of the candidate species, type of natural foods, type of feeding habits (herbivore, detritivore, omnivore or carnivore), feeding stratum (bottom, column, and surface), visual or olfactory feeder, rapid or slow feeder.
- ☆ Besides, other criteria are preferred physical texture, method of feeding (swallowing or nibbling), anatomy of mouth parts and organs of prehension.

B. Information on Ingredients

It includes availability of ingredients, nutrient composition, cost, seasonal, geographical variations, state of availability (dry or wet) and processing methods (solvent extracted, mechanical extracted).

C. Processing at Feed Mill

Type of feed preparation: Mash, crumbles, pellet, flakes. sinking or floating pellet, and type of machinery used.

Based on these informations, binder addition, heat labile additives addition in feed formulation is determined.

Approaches for Feed Formulation

A. Pearson Square

It is one of the earliest methods used in feed formulation. This method facilitates the calculations involved in blending of two ingredients or products with different nutrient concentrations to yield a product with a specified nutrient concentration. Until the advent of computers, this method was a mainstay in feed formulation. In the following example, the target is to produce a 20 per cent crude protein (CP) product by blending two ingredients, a 38 per cent CP ground nut oil cake (GNC) and a 12 per cent CP wheat (Houser and Akiyama, 1997).

1. Place the per cent CP of the GNC in the left hand corner.
2. Place the per cent CP of the wheat in the lower left hand corner.
3. Place the desired per cent CP in the middle.
4. Subtract the smaller number from the larger number diagonally in this case, 38 minus 20 and 20 minus 12 the results, parts of each ingredient that are required in the formulation, are noted in the right corner on the same horizontal line as the ingredient.
5. Add the total of the two numbers on the right hand side.
6. Divide the parts for each ingredient by the total parts and multiply by 100 to determine the percent of each ingredient needed.

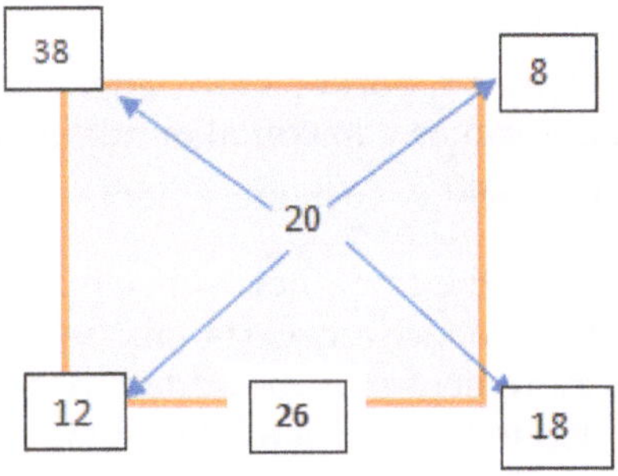

Figure 9.1. Pearson Square.

8/26=30.8 per cent GNC; 18/26=69.2 per cent wheat

Proof: 30.8 x 0.38 = 11.70

69.2 x 0.12 = 8.30

Total 20.00

If more ingredients are to be included, the above product which is actually a blend of two ingredients may be combined with a third ingredient in a second Pierson Square to produce a blend of 3 products. Extensive information on the use of multiple Pierson Squares in feed formulation is reviewed by Church and Nipper 1984.

B Linear Programming/Least Cost Formulation

In aquaculture feed, the number of possible ingredient sources and the number of nutrients present in each, for which requirements have been estimated, are extensive. Thus, a large number of arithmetic calculations are required to produce a least-cost diet. Although 'hand' formulations using a simple calculator may be sufficient (if not a little tedious and time consuming) for the scientist or farmer wishing to formulate a feed for use within his or her own laboratory or farm, this is not the case for the commercial feed manufacturer where time is money and profits are made or lost in the area of feed formulation as the raw material cost is currently representing 80 per cent percent of finished feed (NRC 2011).

To meet this demand, the animal feed manufacturing industry has been employing the computer technique of linear programming since the mid-nineteen fifties. Linear programming is essentially a mathematical tool by which resources are evaluated or selected to achieve an optimal solution to a problem. The value of linear programming is in the number of ingredients and number of requirements or restrictions which can be handled in a short period of time.

Certain data must be supplied to the computer together with the programme to be executed. These data are:

- ✰ The detailed feeding standard which has to be satisfied (*i.e.* dietary nutrient requirements), together with any permitted deviations in each nutrient;

- ☆ The detailed nutrient composition of each potential ingredient;
- ☆ Any restriction on the proportion of the final mix that any one ingredient may represent, this can be a maximal or minimal value (constraints)
- ☆ The cost per unit weight of each ingredient.

The formula calculated by the computer will be that meeting the specification at the least cost of ingredients, hence any extra mixing or handling charge due to the inclusion of a certain ingredient must be added to the cost per unit weight of that ingredient or else added to the cost of the formula.

When using a computer it must be borne in mind that the adequacy of a diet compounded to the resultant formula will be affected by two main factors: the extent to which the feeding standards (ie. nutrient requirements) adopted adequately represent the biological needs of the fish or shrimp and the accuracy with which the amounts of nutrients in the ingredients available to the fish or shrimp are known.

Farm Made Feed

The main objective or aim of feed formulation is development of a nutritionally balanced mixture of feed stuffs which will be eaten in adequate amounts to provide optimum growth of the cultured fish at an acceptable cost by utilizing knowledge of nutrient requirements, locally available feed ingredients and digestive capacity of the fish. Before proceeding with formulating a feed, the ingredients are to be selected from available sources. No single ingredient can be expected to provide the entire nutrient requirement. Each ingredient in the diet should be included for a specific reason *i.e.*, either to supply a specific nutrient or physical property to the diet. Seasonal availability of ingredients, proximate composition, digestibility and nutrient availability should be taken into account during selection of ingredients. Constraints of using some feed ingredients having anti-nutritional factors must also be kept in mind. Additional costs for ingredient handling, processing prior to mixing or pelletizing and transportation cost are also to be considered (Boonyaratpalin and New, 1993)

Formulation of a feed by the nutritionist is only the beginning of a process that ends when the feed is finally consumed. Feed formulation is essentially a recipe making process keeping in mind, the nutritional requirement of particular species, palatability and growth promoting ability of that feed. These objectives can be achieved by judicious selection of feed ingredients, mixing them in proper proportion and presenting them in a most acceptable form (Syed Ahamad Ali, 2018)

Ingredients Used for Farm Made Feed

There are 3 factors to consider in the choice of ingredients:

a) Quality - nutrient composition and presence of any anti-nutrient substances that interfere directly with the absorption of nutrients or contaminants.

b) Quantity – how much is available locally and is the supply regular?

c) Price or cost

Special care should be taken to inspect the materials for wetness, mould growth, insects and parasites. The inspection procedure starts with visual examination of a sample for the colour and texture and smelling for obvious contamination or rancidity (Halver and Tiews, 1979)

Following ingredients may be used for feed of Marine fish

Plant sources: Wheat flour, rice flour, maize flour, brewery wastes, ground nut oil cake, mustard oil cake, soybean meal, sunflower cake, sesame cake (FAO, 1978).

Animal sources: Fish meal, poultry by product and offal, blood meal, bone meal, meat meal, shrimp meal, clam meat, mussel meat *etc.*

Farm Made Pellet Feed Preparation

For preparation of farm made feed the following process are to be carried out:

STEP 1-Grinding

Finely ground ingredients of uniform particle size are to be chosen since dissimilar sizes results in an unstable pellet. Individual ingredients should be ground using a hammer mill, or other type of grinder or even a mortar and pestle. A sieve can also be used to remove large particles or foreign material like stones, pieces of metal *etc.* that can damage machinery.

STEP 2-Weighing

Weigh or measure the feed ingredients as per the formula. Take particular care when weighing micro-nutrients (vitamin and mineral premixes) as these are used in very small quantities and are very expensive.

Figure 9.2. Weighing.

STEP 3-Mixing

Mix all the ground ingredients thoroughly by hand except vitamin and mineral mixture. If large batches are to be prepared, the dry ingredients can be mixed in a horizontal mixer or even in a cement mixer. Poor mixing will result in variation of daily nutrient intake. Good mixing can also improve palatability.

STEP 4-Mixing of Oil

Add the oil and then mix for at least another five minutes. To ensure oil is well mixed throughout the ingredients, it is good to warm the oil or make an emulsion with warm water and mixed with dry ingredients slowly.

STEP 5-Addition of Water

Add water and mix well to form a mash with a cake-like consistency. Water should be added slowly and small test batches of the mixture extruded through the pellet machine (mincer) to see how easily the mixture passes through the die and how the pellets hold together. As a general rule, the total moisture content of the mash should be in the range of 45 to 55 per cent to produce good pellets. If moist ingredients like trash fish, minced poultry *etc.* are used, less moisture will be needed (*e.g.* 25-45 per cent). Adjustments must also be made depending on the type of binders, if any, are used. Binder is not required in farm made feed of Asian seabass as they take the feed immediately after offering.

STEP 6-Steam Cooking

Cook the mixture in an idly cooker or in a big container at 100 °C temperatures for 5-10 minutes. Take out and cool the feed mixture. Steam cooking will gelatinize the starch and also sterilize the feed mixture.

STEP 7-Incorporation of Vitamin-Mineral Mixture

Mix vitamins and minerals with small amount (*e.g.* 10 per cent of total batch) first then blend into larger mixture and ensure that the vitamins and minerals are evenly distributed within the mixture. Vitamin and mineral mixture may be added with dry ground ingredients before steam cooking, but, in that case heat labile vitamin will be destroyed during cooking.

STEP 8-Pelletisation

Pass the feed mash mixture through a pellet machine (mincer) with a 1, 2, or 3 mm die depending on the size of the fish that is being fed. Cut the pellet (which look like noodles) into similar lengths to the closest pellet diameter (*i.e.* 2 mm long for 2 mm diameter pellets for fish). Pellets can be cut off with a knife during extrusion or broken into smaller lengths after they have been dried.

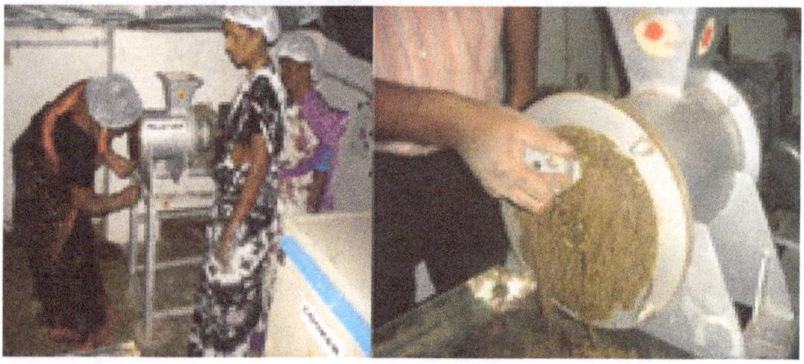

Figure 9.3. Pelletisation.

STEP 9-Drying

The moist pellets should be dried to a moisture content of 10 per cent or less. Ideally, this should be at low temperature (less than 60 °C) and with good airflow to dry the pellets as quickly as possible to ensure that heat-sensitive micronutrients such as vitamins are not destroyed. This can be achieved using:

- ☆ A fan-forced oven (*e.g.* set at <60 °C) for several hours,
- ☆ A simple drying cabinet (with hot air supplied by a heater blower),
- ☆ A solar dryer or
- ☆ Simply by spreading the pellets in the sun.

It is most important to prevent fungal contamination and to avoid an excessive loss of critical nutrients when drying the pellets. Fungus can be toxic to fish and to the humans who handle the feed.

Step 10-Storage

When pellets are dry and cool they should be stored in bags or containers that can be sealed against insects, rats or other pests and to keep out moisture. Avoid using plastic bags because feeds can sweat and this encourages growth of mould. Farm made feed should not be stored for long time. Freshly prepared farm made feed should be fed to fish immediately.

Advantage of Farm Made Feed

- ☆ Feed formulae may be diverse as per availability and cost of ingredient
- ☆ Feed cost can be minimized
- ☆ Farmers can prepare the feed at their farm itself and feed their animal a fresh feed.
- ☆ Loss of vitamin and mineral during preparation can be minimized.

Drawbacks of Farm Made Feed

- ☆ Poor water stability of feed which may pollute pond water if not consumed by fish immediately after offering.
- ☆ Quality of raw material varies hence nutrient composition of feed will vary.
- ☆ Digestibility of farm made feed may be inconsistent.
- ☆ Large scale aquaculture with farm made feed is not feasible

References

1. Boonyaratpalin, M. and M.B. New. 1993. On-farm feed preparation and feeding strategies for marine shrimp and freshwater prawns, p. 120-134. *In* M.B.New, A.G.J. Tacon and I. Csavas (eds.) Farm-made aquafeeds.Proceedings of the FAO/AADCP Regional Expert Consultation on Farm-Made Aquafeeds.14-18 December 1992, Bangkok, Thailand.FAO-RAPA/AADCP, Bangkok, Thailand, 434 p.
2. Church, D.C. and W.A. Nipper. 1984. Ration formulation. in D.C. Church, (ed). Livestock feeds and feeding, pp 195-227
3. De Silva, S.S. 1989. Reducing feed costs in semi-intensive aquaculture systems in the tropics. Naga, The ICLARM Quarterly 12(4):6p

4. FAO.1978 Fish Feed Technology. In: Aquaculture development and Co ordination Programme. ADCP/REP/80/11 p. 289 325.
5. Halver, J.E. and Tiews, K. 1979. Finfish nutrition and fish feed technology. Vol. II 613 p.
6. Houser R.H. and Akiyama D.M., 1997. Feed formulation principles. In: Crustacean Nutrition, L.R. D'Abramo, D.E. Conklin and D.M. *Akiyama* (eds), pp. 492–518
7. National research council.2011. Nutrient Requirement of fish and shrimp. Washington, DC. The National Academic press. https://10.17226/13039
8. Syed Ahamad Ali.2018. Nutritional Feeding of Fish and Shrimps in India. Om Publictions, N.Delhi

Chapter 10

Live Feed Culture: Micro Algal Production and Preparation of Algal Concentrates

Biji Xavier, Ritesh Ranjan, Sekar Megarajan and Shubhadeep Ghosh

Visakhapatnam Regional Centre of Central Institute of Marine Fisheries Research Institute (CMFRI), Visakhapatinam, Andhra Pradesh

Phytoplankton species play a crucial role in the food chain as a key nutrient in aquatic environments. These species use ammonia, urea, nitrate and phosphate nutrients, vitamins, trace elements and some form of organic matter formed by chemical breakdown of dead cells (Gokpinar, 1990). The marine algae cultures have a great importance in the economy both in the laboratory as well as in the natural environment. Phytoplankton or zooplankton is the first step in the food chain for the most marine fish farming as feedstock. For this reason, proper feeding of marine fish larvae, production of phytoplankton and zooplankton has a great importance. Problems in marine fish farming, as well as the culture of oysters and shrimp solved with the development of single-celled algae production methods.

Microalgae, belonging to various taxonomic groups, are microscopic single celled, multi-celled, planktonic or benthic algae. Microalgae are the first staple of shellfish and fish larvae of aquatic animals. In addition it has a great importance for feeding other fish staples like rotifers, *Cladocera*, or copepods (Timur, 1992). Microalgae can produce not only biomass but also complex molecules necessary for life activities with the assistance of any light source (sun or artificial light); by combining various sources such as nutrient salts and CO_2.

Species Selection for Culture

All microalgal species are not equally suitable for culture in laboratory or hatchery. A suitable species or strain of microalgae should be selected depending upon the following criteria:

1. Species of finfish or shellfish which feed on the microalgae
2. Potential or viability for Mass culture
3. Cell size
4. Digestibility
5. Overall nutritional value

Common Micro Algae and Applications in Aquaculture

Nannochloropsis oculata

It is a small (2-5 μm) green algae belong to the family Eustigmatophyceae, rich in Eicosa Pentaenoic Acid (EPA). It is used for rotifer culture and as water conditioner in finfish hatcheries. It is also used in reef tanks for feeding corals and other filter feeders.

Isochrysis galbana

These are small (3-5 μm) golden brown flagellates belonging to the family Isochrysidaceae. They are excellent feed for the production of copepods, brine shrimp, oysters, clams, mussels and scallops. Because of its richness in Docosa Hexaenoic Acid (DHA), *Isochrysis* (Figure 10.1) is used for the enrichment of zooplankton (copepod, *Artemia*) which has been raised on *Nannochloropsis* (Figure 10.2) and yeast, for improving the EPA/DHA ratio in the live feed.

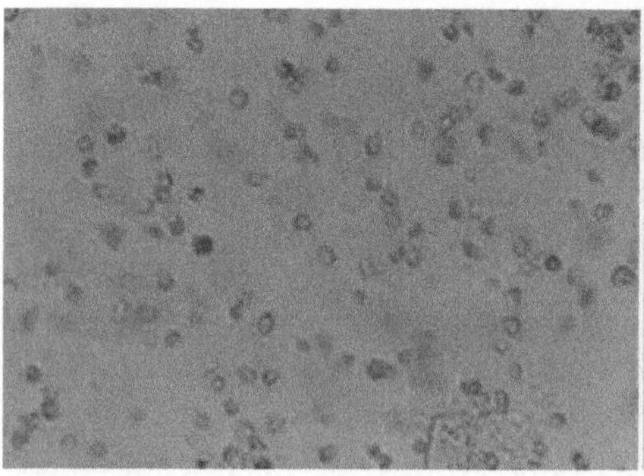

Figure 10.1. ***Isochrysis galbana***

Figure 10.2. ***Nannochloropsis oculata***

Chaetoceros calcitrans

It is a marine planktonic diatom belonging to the family Chaetocerotaceae

(Figure 10.3). The cells (*diameter* 2 - 85 µm, *length* 2 - 45 µm) are elliptical in valve view. It is the widely used species in aquaculture, as it is composed of nutritional value suitable for marine filter feeders especially molluscs. It is used in shrimp hatcheries to increase the vitamin levels.

Chlorella salina

It is single celled green algae, spherical in shape, 2-10 µm in diameter. Used for feeding rotifers in finfish larval rearing (Figure 10.4).

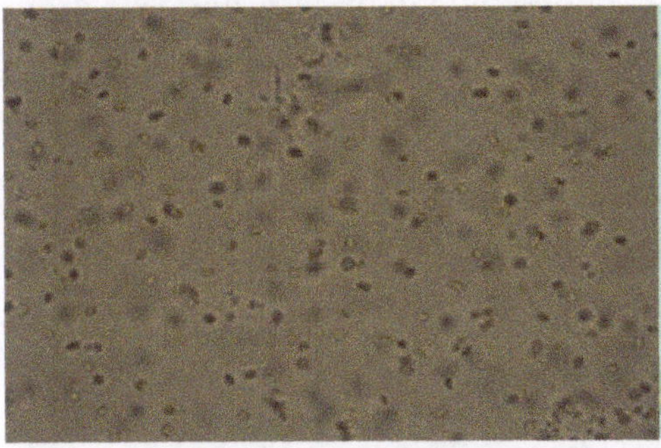

Figure 10.3. ***Chaetoceros calcitrans***

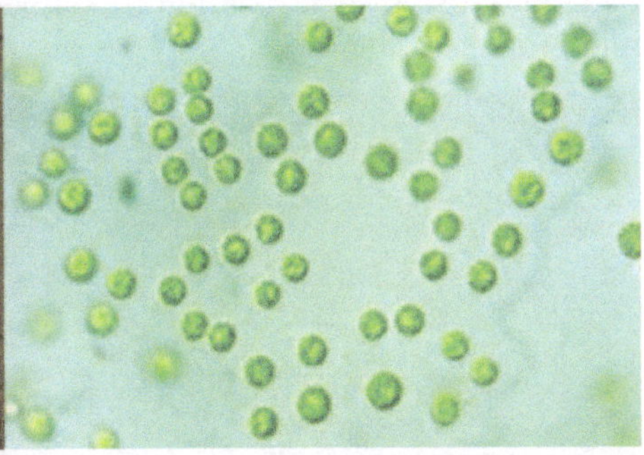

Figure 10.4. ***Chlorella salina*** **(*Source*: Google)**

Pavlova lutheri

It is a small (3-10 µm) golden brown flagellate, belong to the family Haptophyceae (Figure 10.5). It is a widely cultured species and used in the cultures of oysters, clams, mussels and scallops to increase the DHA/EPA levels in their broodstock. It is a temperate species, used for rotifer enrichment, which is mainly used in cold water fish hatcheries (cod).

Figure 10.5. ***Pavlova lutheri***

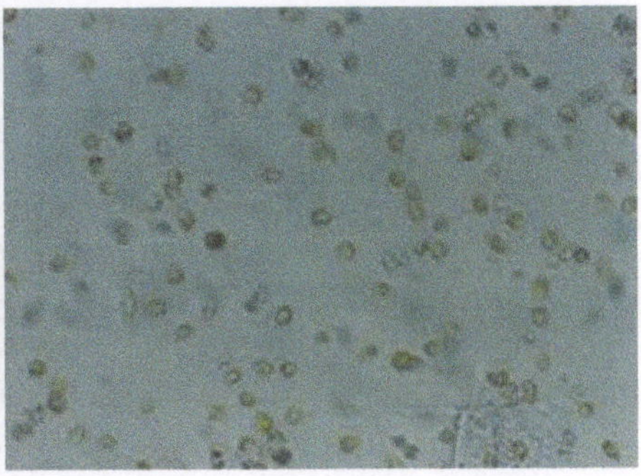

Figure 10.6. ***Tetraselmis suecica***

Tetraselmis suecica

It is a large green (4-5 µm) flagellate belonging to the family Chlorodendraceae (Figure 10.6). This microalgae forms the standard feed for oysters, clams, mussels and scallops. This particular algae is rich source of natural amino acids that stimulate feeding in marine animals and also forms excellent feed for shrimp larvae. It is used in combination with *Nannochloropsis* sp. for rotifer production and also for feeding the brine shrimp.

Thalassiosira weissflogii

It is a large diatom (4-32 μm) cylindrical in shape belonging to family Thalassiosiraceae (Figure 10.7). It is the major species used in shrimp and shellfish larviculture. It is the single best algae for shrimp larvae and also good feed for copepods, brine shrimp, and broodstock conditioning of oysters (post set), clams and mussels.

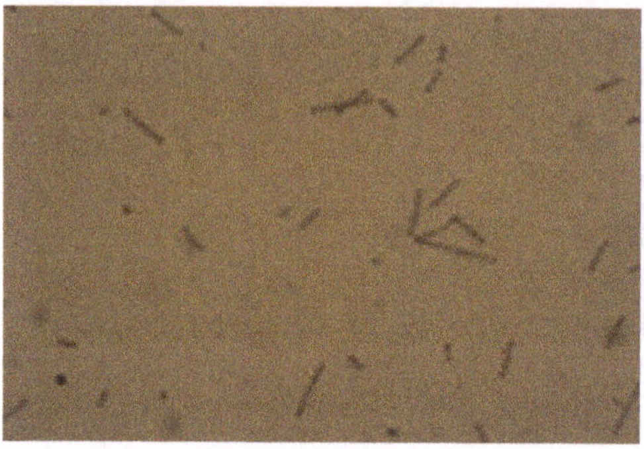

Figure 10.7. ***Thalassiosira weissflogii***

Figure 10.8. ***Skeletonema costatum***

Skeletonema costatum

These are diatoms belongs to the family Skeletonemaceae (Figure 10.8). Cells are short (2-21 μm) and cylindrical appearing as long straight chains. It is widely used for extensive and intensive shrimp hatchery systems.

Microalgae Culture

Microalgae culture is a form of aquaculture involving farming of different species of microalgae in a confined environment. The culture has three components namely, culture medium contained in a suitable vessel; the algal cells growing in the medium and air, to allow exchange of carbon dioxide between medium and atmosphere. An autotrophic alga need light, CO_2, water, nutrients and trace elements for their growth (Table 10.1). Some auxotrophic algae, which cannot prepare biochemical compounds like vitamins through photosynthesis, need additional biochemical compounds to be added in the culture medium.

Growth Dynamics of Microalgae

The knowledge on microalgae growth dynamics is important for aquaculturist to know when to harvest the microalgae, to estimate growth rate and population doubling time *etc.* Generally, the microalgal culture follows a characteristic pattern of growth which includes five reasonably well defined phase of growth in batch culture (Fogg and Thake, 1987). (i) Lag or Induction phase (ii) Exponential phase (iii) Declining phase (iv) Stationary Phase (v) Death phase

Table 10.1. Temperature, Light and Salinity Range for Culturing Selected Microalgae Species

Species	*Temperature (ºC)*	*Light (Lux)*	*Salinity (ppt – ‰)*
Chaetoceros calcitrans	25 - 30	2,000-10,000	20 – 35
Isochrysis galbana	25 -30	2,000-10,000	10 – 30
Skeletonema costatum	10 - 27	2,500-5,000	15 – 30
Nannochloropsis oculata	20 - 30	2,500-8,000	12 – 30
Pavlova sp.	15 -30	4,000-8,000	10 – 40
Tetraselmis sp.	20 -28	5,000-10,000	20 – 40
Chlorella sp.	10 -28	2,500-5,000	26 – 30
Thalassiosira sp.	25 - 30	2,000-10,000	20 – 35

i. **Lag or induction phase:** The Inoculation of culture into a new medium is acclimatized with the surroundings or to the new physico - chemical conditions during which there will not be any cell division for slight time thus the stage is known as lag or induction phase.

ii. **Exponential phase:** Once the inoculum is acclimatized to the new physico-chemical conditions, it starts multiplication and grows rapidly, thus this phase is known as exponential phase. The duration of exponential phase in cultures depends upon the volume of the inoculum, the growth rate and the capacity of the medium and culturing conditions to support algal growth.

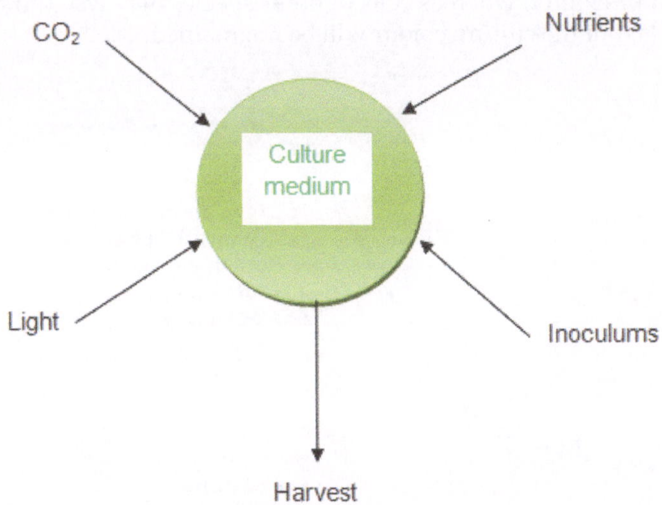

Figure 10.9. Process of Micro Algal Culture with Various Inputs.

iii. **Declining phase:** When the death of the cells will exceed the multiplication of the cells in a culture, then the phase is known as declining phase. This phase generally occurs in culture when either a specific requirement for cell division is limiting or something else is inhibiting reproduction. The biomass is often very high and exhaustion of a nutrient salt, limiting carbon dioxide or light limitation becomes the primary cause of declining growth. At low cell densities too much CO_2 may lower the pH and depress the growth. CO_2 limitation at high cell densities control any further biomass increase to be linear rather than exponential (with respect to time) and proportional to the input of CO_2. Light limitation at high biomass occurs when the cells absorb most of the incoming irradiation and individual cells shade each other. This is known as self-shading.

iv. **Stationary phase:** After the arrested growth, the culture will be stationary without any further cell division. The net growth of the culture becomes zero during this phase and within an hour's culture cells may undergo dramatic biochemical changes. The nature of the changes depends upon the growth limiting factor. Nitrogen limitation may result in the reduction in protein content and relative or absolute changes in lipid and carbohydrate content. Light limitation might affect the pigment content of most species and changes in fatty acid composition. The longer the cells are held in this condition the longer the lag phase will be when cells are returned to good growth conditions for sub culture.

v. **Death phase**: This is last phase of growth dynamics, when the cells may lose its viability and start to die. The death phase of culture is generally very rapid, thus the term 'culture crash" is often used for this phase. The cultures of some species will loose their pigmentation and appear washed out or cloudy, whereas cells of other species may lyse (no recognizable cells) but the culture colour will be maintained.

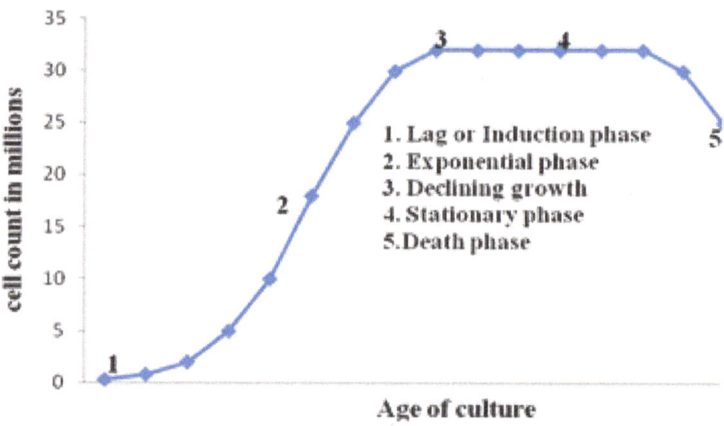

Figure 10.10. Growth Curve of Micro Algae (*Nannochloropsis* sp.).

Algal Culture Techniques

The following are the different algal culture techniques used for microalgae culture.

Batch culture: In this culture the resources present in the culture medium are abundant; the micro algae grow according to sigmoid curve. Once the resources are utilized by the cells, sub culturing is followed by transferring a small volume of the existing culture to the large volume of fresh culturing medium.

Continuous culture: In this culture, the resources are potentially infinite. Culture is maintained on chosen point on the growth curve by the regulated addition of fresh medium.

Indoor/outdoor: Indoor culture allows the control over illumination, temperature, nutrient level, contamination with predator and other competing algae; however, outdoor culture will not have any control on the above said parameters.

Open/closed: Open cultures in uncovered tanks and ponds are more contaminated than closed cultures in test tube, conical flask *etc.*

Axenic (Sterile)/Xenic

Axenic cultures are pure cultures free from all foreign contaminants and this type of culture requires strict sterilization of glass wares, culture media and vessels to avoid contamination, where as xenic culture might be a contaminated one.

Microalgae Culture Media

Algal nutrient solutions or culture medium are made up of a mixture of chemical salts and water. The culture medium provides the nutrient needed for the growth of algae. These nutrient solutions are formulated specifically for its use in aquatic environments and their consistency is more precise for laboratory culture. The selection of culture media mainly depends on the type algae species cultured. Most commonly used culture media used for stock culture and mass culture of micro algae in the laboratory is 'Conway' or 'Walne's medium (Walne,1974) (Table 10.2) and Guillard's F/2 media (Table 10.3). This media is mainly used for indoor culture of *Nannochloropsis*, *Chlorella*, and diatoms like *Chaetoceros*, *Skeletonema*, *Thalassiosera* and *Tetraselmis*. The culture medium is constituted with the addition of macronutrients (nitrates, phosphates and silicates), micro nutrients (Fe, Mn, Co, Zn, Cu and Ni) and vitamins (B_1, B_{12} and B_7).

Table 10.2. Composition of Conway Medium

	Solution (A)	
1.	Potassium Nitrate (KNO_3)	100gm
2.	Sodium di-hydrogen orthophosphate ($NaH_2PO_4.2H_2O$)	20gm
3.	EDTA di-sodium salt (Na_2EDTA)	45gm
4.	Boric Acid (H_3BO_3)	33.4gm
5.	Ferric Chloride ($FeCl_3$)	1.3gm
6.	Manganous Chloride ($MnCl_2.2H_2O$)	0.36gm
7.	Distilled Water	1lit

	Solution (B)	
1.	Zinc Chloride ($ZnCl_2$)	4.2gm
2.	Cobalt Chloride ($CoCl_2.6H_2O$)	4.0gm
3.	Copper Sulphate ($CuSO_4.5H_2O$)	4.0gm
4.	Ammonium molybdate ($(NH_4)_6Mo_7O_{24}.4H_2O$)	1.8gm
5.	Distilled Water	1lit
6.	Concentrated HCl	
	Solution (C)	
1.	Vitamin B_1 (Thiamine)	2gm
2.	Vitamin B_{12} (Cyanocobalamin)	100mg
3.	Distilled Water	1lit
	Solution (D)	
1.	Sodium silicate ($Na_2SiO_3.5H_2O$)	40ml
2.	Distilled water	1lit

Stock culture: Only autoclaved seawater should be used. Working Solution for mass culture: Add 1ml of Solution A, 0.5 ml of B, 0.1 ml of C and 1ml of D into 1 lit sea water. D: Only for Diatoms (*Chaetoceros*)

Table 10.3. Composition of Guillard's F/2 Media

	Solution (A)		
1	Sodium nitrate ($NaNO_3$)	75g	1 Lt distilled water
2.	Sodium phosphate($NaH_2PO_4.H_2O$)	5g	
	P.S.Solution		
1	Copper Sulphate ($CuSO_4.5H_2O$)	10g	1 Lt distilled water
2	Zinc Sulphate ($ZnSO_4.7H_2O$)	22g	1 Lt distilled water
3	Cobalt Chloride($CoCl_2.6H_2O$)	10g	1 Lt distilled water
4.	Manganous Chloride($MnCl_2.4H_2O$)	180g	1 Lt distilled water
5.	Sodium Molybdate ($Na_2MoO_4.2H_2O$)	6g	1 Lt distilled water
	Prepare each solution separately in 1lit bottles		
	*Solution (B)**		
1.	EDTA di-sodium salt (Na_2EDTA)	4.36g	1 Lt distilled water
2	Ferric Chloride ($FeCl_3.6H_2O$)	3.15g	
	Add 1ml of each solution (P.S solution1-5) each to 1 lit of EDTA and $FeCl_3$ mixed solution		
	*Solution (C)***		
1	Thiamin HCl	20g	1 Lt distilled water
2	Biotin	100mg	1 Lt distilled water
3	Cyanocobalamine (B_{12})	100mg	1 Lt distilled water
	Add 5ml of each solution into 1lit of sea water		

	*Solution (D)****		
1.	Sodium Silicate ($Na_2SiO_3.9H_2O$)	35g	1 Lt distilled water

Stock culture: Only autoclaved seawater should be used.

*Solution B: 1lit of EDTA and $FeCl_3$ mixed solution with P.S. solution (1ml of each)

** Solution C: 1lit of sea water with Thiamin, Biotin and Cyanacobalamin solution (5ml each)

*** Solution D: Only for Diatoms (*Chaetoceros*)

Working Solution for mass culture: Add 1ml of Solution A, B, C and D into 1 lit sea water. D: Only for *Chaetoceros.*

Stock Solutions Preparation

Stock solutions are prepared with accurate weighing of the chemicals in the specific culture media, dissolved in the specific volume of distilled water. There are two terms used in stock solution preparation, working stock and primary stocks. **Working stocks** are the small quantity (aliquot) of solution which is directly used for the preparation of final medium so that adding a small volume of liquid stock solution is easier and quicker than weighing of individual dry chemicals. **Primary stocks** are formed from several single substance solutions and finally combined to form the working stock. Stock solutions made from chemicals like EDTA, *etc.*, and it may need heat treatment to dissolve completely in the water to avoid unnecessary precipitation of the nutrients in the medium. But, vitamin stocks should be prepared with normal distilled water and should not be exposed to any heat treatments and are advised to be kept in dark bottles.

Isolation of Microalgae

Collection of micro algal sample is the crucial step which can be done using plankton nets of mesh size 10, 30 and 120 µm. The collected micro algae need to be kept in sterile nutrient solution to keep them alive and allow them to grow. During sample collection both biotic (light, water temperature, dissolved oxygen, dissolved carbon dioxide, nutrient concentration, pH, salinity) and abiotic factors (pathogen) at the sampling site need to be considered. Technique used for the isolation of micro algae includes technique of dilution, single cell isolation by micropipette and agar streaking.

i. Dilution Technique

This isolation technique is suitable for organisms which are abundant in samples and ineffective for rare organisms (Figure 10.11). This technique enables isolation of a single cell by repeated attempts. In dilution technique one inoculates large number of test tubes with culture media (dilution ranging from 10^{-1}-10^{-10}). After inoculating, incubate the test tubes under controlled temperature and light. Examine the culture after 2-4 weeks by taking a small sample aseptically from each dilution tube. Uni-algal culture may grow in any of the higher dilution tubes. The success of the present technique is highly dependent on the accuracy of a measured amount

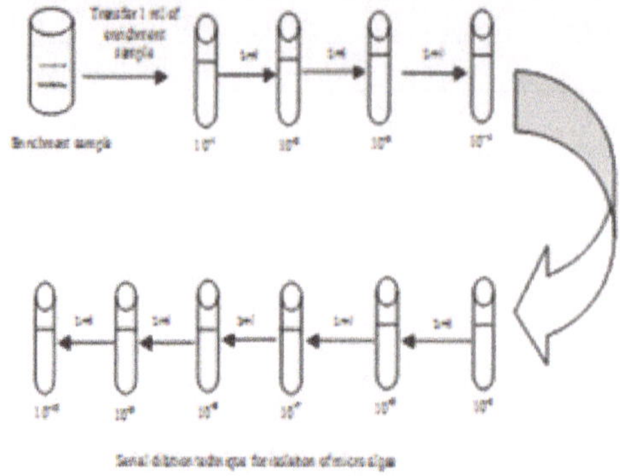

Figure 10.11. Serial Dilution Technique for Isolation of Micro Algae.

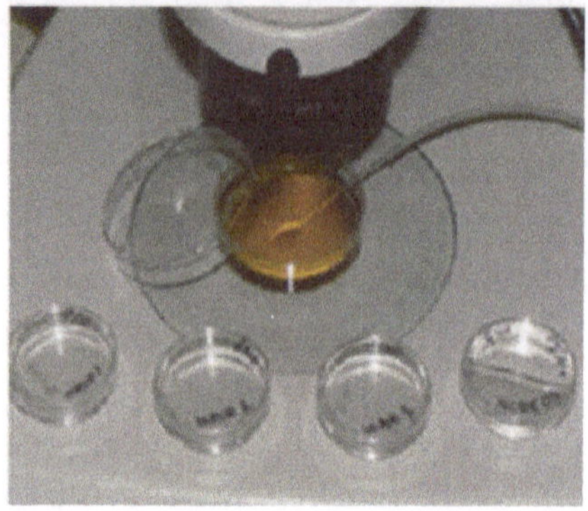

Figure 10.12. Isolation of Single Cell with the Help of Microscope.

of cell culture during transfer from one medium to another. Axenic isolates are not possible with dilution technique because bacteria are more abundant than algae.

ii. Single Cell Isolation by Micropipette

This is the most common isolation technique, which is performed with a pasture pipette or glass capillary under microscopic observation (Figure 10.12). These single cells are transferred to sterile droplets of water or suitable media. The technique requires expertise and precision. The cells can get damaged due to shear stress caused by micropipette or capillary tips. Hence, caution is required while doing this technique. Ultrapure droplets are required, especially for marine samples, to distinguish between microalgal cells and other particulate matters.

iii. Streaking Cells along Agar Plates

For preparing the agar medium, 0.8 to 2 gm of agar is added to a litre of natural filtered seawater in a conical flask. The flask containing agar is usually heated on a flame and boiled twice till the agar is dissolved in water. The culture media 'or' nutrients (solution A and B of Conway media or F/2 media) are added to the agar solution. The flask mouth is usually covered with aluminium foil and then the flask is autoclaved for 15 minutes under 15 lbs pressure and 121°C temperature. The petri dishes are usually sterilized by keeping them in hot air oven at 150°C for 30 minutes. The solution C of culture media is added aseptically to the autoclaved agar flask once the temperature has been brought down to 40°C. Then, agar plates are prepared aseptically by pouring the warm autoclaved agar into the sterile petri dishes near a Bunsen flame or in a laminar flow, followed by covering up the petri dishes and leaving them to cool for about 2 h.

Isolation is accomplished by streaking the sample across the agar surface (0.8-2 per cent). The streaking can be done with the help of 'loop' or spreader (Figure 10.13). After streaking, the agar plate is incubated until colonies of the cells appear. The isolated colonies are removed from the agar plate and can be further subcultured onto other agar plates. Axenic cultures without contamination can be produced with this method.

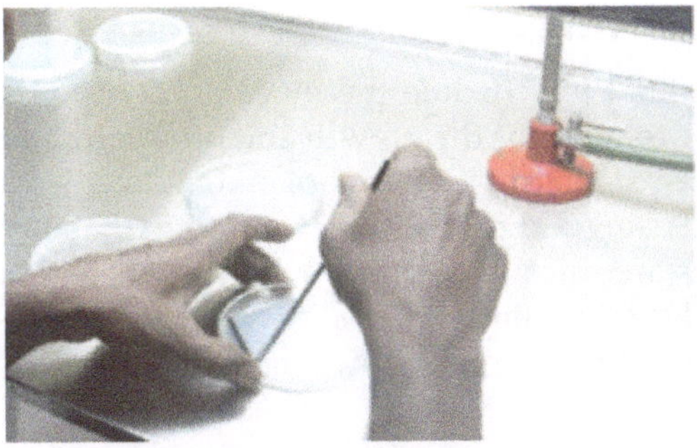

Figure 10.13a. Steaking of Micro Algal Culture on Agar Plates.

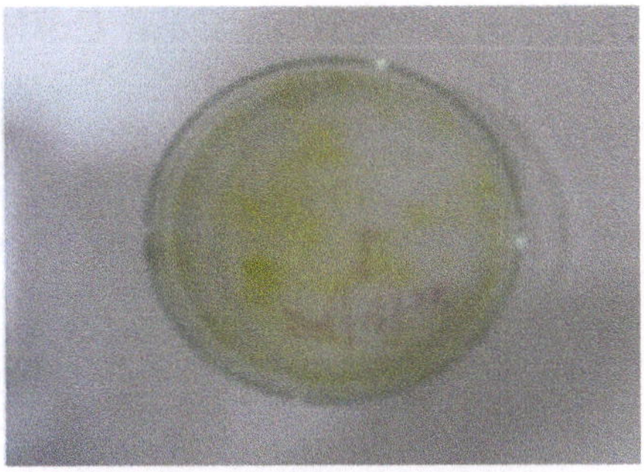

Figure 10.13b. Micro Algae (*Nannochloropsis* sp.) Colonies on in Agar Plates.

Maintenance of Algae Culture

Algal multiplication follows various steps like culture, water sterilization, nutrient enrichment, inoculation of new culture with the pure algal strain, microscopic observations for growth of the algal cells and finally mass culture of algae in larger containers.

Stock Culture

It provides reservoir of algae cells from which large cultures can be initiated. Stock cultures are kept in small volume containers.

Subculture

Subculture involves inoculating some cells from old stock culture to fresh culture medium, so that the cells can continue to grow, divide and be healthy. If sub culturing is not done, cells in the stock culture will die eventually. Precaution must be taken while transferring the stock culture, so that contaminants from air can not enter into the stock culture.

Figure 10.14. Stock and Intermediate Culture of *Nannochloropsis oculata* and *Isochrysis galbana*.

Determination of Algal Cell Densities

Regular algal cell counting is essential to schedule the inoculation of mass culture from the containers. It is mainly to monitor the growth and quantification of the algae to be fed to the larval organisms. Most of the nannoplankton organisms are less than 10 µm and they are counted using Sedgwick Rafter Counting Chamber or Haemocytometer. Haemocytometers are thick glass slides with two chambers on the upper surface, each measuring 1.0 x 1.0 mm. A special cover slip is placed over these two chambers giving a depth of 0.1 mm making the total volume of each chamber 0.1 mm^3.

Sample is prepared and treated with eosin or formalin to kill the cells. After placing the cover slip on the Haemocytometer, pipette with sample should be brought to the edge of the Haemocytometer and touch it to release the sample. The sample will run inside and form a thin film of the culture in which the cells will be uniformly distributed. Haemocytometer has got 9 chambers (4 sides having 16 divisions and 5 chambers of multiple divisions). The counting of algal cells can be carried out in the middle chamber, which is consisting of 5 x 5 cells. Again the each cell has 16 divisions (4 x 4). In this division, the presence of microalgal cells are counted in 4 corner and center cells and then the following formula is applied:

Total number of cells/ml = (Average Number of cells from 5 square) x 25 x 10^4

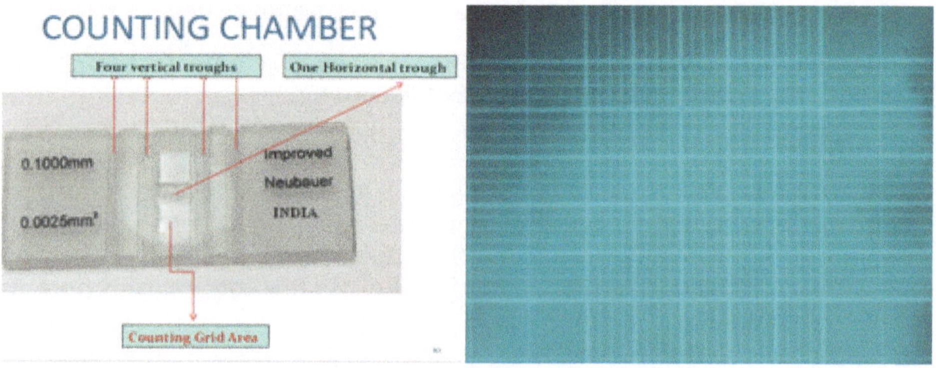

Figure 10.15. Algal Cell Counting Meter.

Mass Culture of Microalgae

Since large quantity of micro algae is needed for zooplankton as well as fish larval rearing system, the cultures need to be multiplied within the limited period of time. Hence, mass culture is to be done with suitable culture medium to meet the algal requirement. Mass culture can be carried out in 10 lit plastic buckets, 20 lit pet bottle, 100 or 200 lit perspex and glass tanks. This mass culture can be of batch culture, semi-continuous or continuous culture. A small inoculum equal to 0.5 per cent of the volume of the new culture will normally generate new healthy cultures for healthy cells of a robust species. However, if the species is delicate or the culture is less healthy, then a larger inoculum of 10 per cent - 20 per cent may be needed to support a new culture. For the mass culture of micro algae media named TMRL (Tung Kang Marine Res. Lab) is reported to be effective (Table 10.4). However, agricultural chemicals (Table 10.5) are used generally for mass culture of microalgae in order to reduce the cost of production.

Table 10.4. Composition of TMRL Medium (100 ml)

Potassium nitrate	10 gm
Sodium orthophosphate	1 gm
Ferric Chloride	0.3 gm
Sodium Silicate	0.1 gm

Table 10.5. Commercial Fertilizer Medium (1t)

Ammonium sulphate	100 gm
Urea	10 gm
Tripple super phosphate	10 gm

Development of maximum density of algal cells in outdoor culture depends mainly on factors like atmospheric light and temperature, salinity of sea water without any contaminants and a good density of indoor inoculums. In outdoor culture with 100 l tanks around 10-15 million cells per ml can be achieved within 2 days of inoculation and around 3-5 million cells per ml can be achieved in 1 ton outdoor tanks with proper environmental conditions.

Micro-Algal Harvest and Concentration

The culture should be harvested during the exponential phase of the micro-algae after determining the cell concentration. If the culture has entered the declining or stationary phase, the metabolites will be very high and the cells may not be in healthy condition. The rearing larval organisms may not show the expected growth if fed with this feed. The trouble which comes frequently like air pressure, salinity, pH, culture contamination and nutrient media for culturing the algae needs immediate attention. Frequent observation to avoid contamination is essential as the algal feed

Figure 10.16. Outdoor Mass Culture of Micro Algae.

Figure 10.17. Outdoor Mass Culture of Micro Algae in 1t FRP Tanks.

is very much important in terms of nutrients for the other organisms like rotifers and copepods to feed on and also pure culture should not be contaminated while adopting the green water technology for the larval rearing system for marine fishes or shell fishes.

Table 10.6. Comparison of Microalgal Harvesting Methods (Christenson *et al.*, 2011; Milledge *et al.*, 2013; Lee *et al.*, 2013)

Method	*Advantages*	*Disadvantages*	*Dry Solids after Harvesting (per cent)*
Centrifugation of microalgae	Cell recovery over 90 per cent, can handle most algal types with rapid efficient cell harvesting, reliable.	High capital and operational costs, energy intensive.	12–22
Filtration of microalgae	Cell recovery 70–90 per cent,wide variety of filter and membrane types available, reliable, can handle delicate cells.	Highly dependent on algal species, best suited to large algal cells, clogging or fouling an issue, high capital and operational costs.	5–27
Sedimentation of microalgae	Cell recovery 10–90 per cent, low cost, potential for use as a first stage to reduce energy input and cost of subsequent stages.	Algal species specific, best suited to dense (heavy) non-motile cells, separation can be slow or unreliable, low final concentration.	0.5–3
Flotation of microalgae	Cell recovery 50–90 per cent,can be more rapid than sedimentation, possibility to combine with gaseous transfer.	Algal species specific, high capital and operational cost, flocculants usually required.	3–6
Flocculation + sedimentation of microalgal flocs	Cell recovery over 90 per cent, wide range of flocculants available, variable price, can be low-cost.	Removal of flocculants, chemical contamination, fragile flocs and/or longer settling times.	3–8
Electroflocculation + sedimentation or flotation of microalgal flocs	Cell recovery over 90 per cent, low energy consumption, possibility to combine with flotation in one step	Contamination of biomass with metal ions, active chlorine can be formed bytreatment of seawater	10 (sedimentation) 30–40 (flotation)
Magnetic separationof microalgae	Cell recovery over 90 per cent, fast, can be considered as one harvesting step, cost effective.	Magnetic modification of biomass required, subsequent obtaining of pure and magnetic particle-free biomass can be problematic.	10–20

Current harvesting methods include biological, chemical, mechanical, and to a lesser extent, electrical operations (Christenson *et al.*, 2011) are used. In general though, there is no proven single best method for harvesting microalgae (Sukenik *et al.*, 1984) as each has its own advantages and disadvantages (Table 10.6). Typical strategies currently applied for harvesting microalgae include centrifugation, filtration, various forms of flocculation (*e.g.* chemical using inorganic and organic agents, alkaline flocculation, bio-flocculation using microorganisms, electro-coagulation), sedimentation, and flotation.

Preparation of *Nannochloropsis oculata* Concentrate

Industrial centrifuge, which can hold more volume of culture, can be used for preparation of the concentrate. The culture is prepared first transferring the culture into centrifugation bottle followed by centrifugation at the maximum speed of 4000 rpm for 30 min. Then the supernatant is decanted and the precipitated concentrate is collected with the help of spatula or any other means without damaging the cells. This method can accumulate a cell count of an approximately 30 billions/ml from centrifugation of 300 litres of *Nannochloropsis oculata* culture.

Preservation in Glycerol

The *Nannochloropsis oculata* concentrate with 5 - 20 per cent inclusion of glycerol is amenable to preservation by both chilling and freezing. However, different concentration of glycerol performs better with chilling (10 per cent) and freezing methods (20 per cent) for the period of 5 months of preservation.

Figure 10.18. Concentrated *Nannochloropsis oculata* Cells before and after Dilution (20X).

Cell Viability Test

Viability of the harvested and preserved cells are tested using 'Evans Blue' stain. Ruptured cells appear blue, since Evans Blue solution diffuses into the protoplasm region and stain the cell blue.

Application of Concentrates in Fish Hatcheries

The preserved *Nannochloropsis* concentrate is to be checked for its efficiency as feed for rotifer, inoculum and green water larval rearing systems.

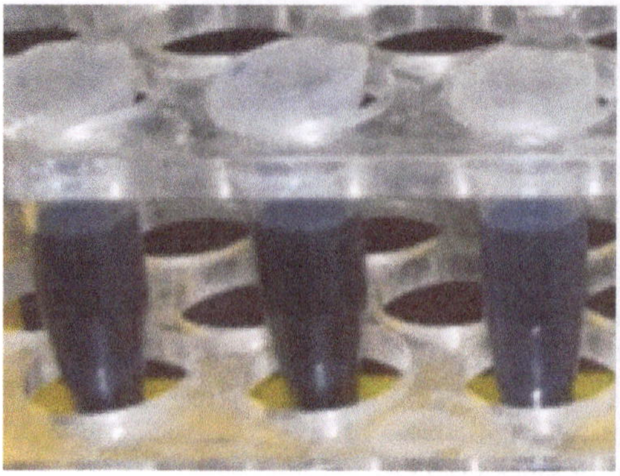

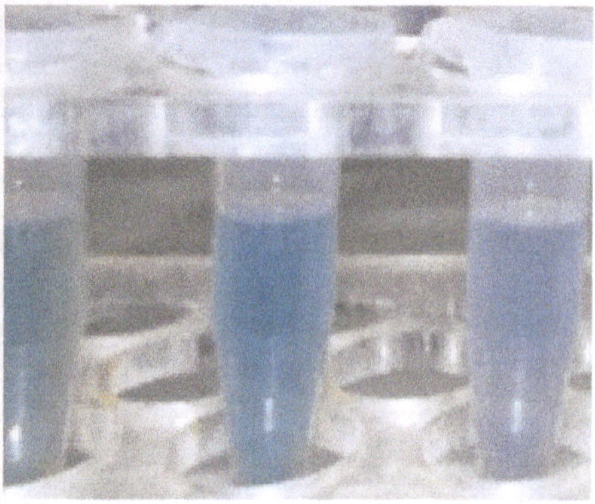

Figure 10.19. Diluted *Nannochloropsis oculata* Cells from Concentrate before and after Evan's Blue Staining.

Nanno Concentrate as Feed for Rotifer Culture

The rotifer (*Brachionus plicatilis*) is cultured using nanno concentrate as feed, and the result concluded that the concentrate persevered with 10 per cent of glycerol helps to achieve maximum of 1040 number of rotifers/ml at the concentration 3x10^6 nanno-cells/ml in rotifer culture.

Inoculum for *Nannochloropsis oculata* Culture

The preserved nanno-concentrates (0 to 20 per cent of glycerol in chilling) are diluted with sterilized sea water to obtain a cell count of 1x10^6/ml. A 10 per cent of inoculum is added into 10 ml of culture medium containing Conway medium. The culture is to be maintained at optimum environmental condition for the stock culture development. In the end of third day of inoculation the, cell count reaches 12x10^6 cells/ml in culture, and the same stock can be used for further development. In another method, the glycerol preserved nanno concentrate is diluted and used as direct inoculum to prepare intermediate culture also.

Green Water System for Larval Rearing

To maintain green water system for marine finfish larval rearing, an approximately 3.5 g of nanno-concentrate is added into 1 cubic meter water volume,

Figure 10.20. Rotifer Culture on *Nannochloropsis oculata* Concentrate.

Figure 10.21. Green Water Rearing System from Nannochloropsis.

and it maintains the nanno concentration of 1 lakh nanno cells/ml in suspended condition, which also maintains the water quality during the period of larval rearing.

References

1. Andersen, R. A., 2005. Algal culturing techniques. Elsevier/Academic press.
2. Christenson, L.; Sims, R.2011 Production and harvesting of microalgae for wastewater treatment, biofuels,and bioproducts. Biotechnol. Adv. 29, 686–702.
3. Coutteau, P. 1996. Micro-Algae. In: Lavens, P., Sorgeloos, P. (Eds.), Manual on the Production and Use of Live Food for Aquaculture. FAO Fisheries Technical Paper, vol. 361. FAO, Rome, pp. 7– 47.
4. Fogg, G.E. and Thake, B. 1987 Algae Cultures and Phytoplankton Ecology. 3rd Edition, The University of Winsconsins Press, Ltd., London.
5. Gopinathan, C.P. 1982. Methods of culturing phytoplankton. In manual of research methods for fish and shell fish nutrition. *CMFRI. Spl. Publ*, **8**:113-118.
6. Gökpýnar, Þ.,1990. Christenson, *et al.*, 2011"Mikroalg Kültürleri I-Uygulama ve Kullaným Alanlarý", Ege Ü. Su Ürünleri Derg., 7; 46-56.
7. Lee, A.K.; Lewis, D.M.; Ashman, P.J. 2013. Harvesting of marine microalgae by electroflocculation: The energetics, plant design, and economics. Appl. Energy 108, 45–53.
8. Maurice Levasseur· Peter A. Thompson And Paul J. Harrison 1993. Physiological acclimation of marine phytoplankton to different nitrogen sources. *Journal of Phycology*. Vol. 29 (5), Issue 5, Pages 587–595
9. Milledge, J.; Heaven, S.2013. A review of the harvesting of micro-algae for biofuel production. Rev. Environ. Sci. Bio-Technol. 12, 165–178.
10. Sukenik, A.; Shelef, G.1984. Algal auto flocculation—Verification and proposed mechanism. Biotechnol. Bioeng. 26, 142–147.
11. Timur, G. 1992. "Plankton Bilgisi ve Plankton Kültürü", Akdeniz Üniv. Yay. No: 40
12. Walne, P,R, 1974. Culture of bivalve molluscs. Fishing News (Books) Ltd. Surrey, G.B., pp. 1-173.

Chapter 11

Broodstock Development and Seed Production of Indian Pompano and Grouper

Ritesh Ranjan, Sekar Megarajan, Biji Xavier, Chinni Bathina and Shubhadeep Ghosh

Visakhapatnam Regional Centre of ICAR-CMFRI

Indian Pompano

The common name or FAO name of *Trachinotus mookalee* is "Indian pompano". *T. mookalee* belongs to the family Carangidae (jacks and pompanos) holds immense

Figure 11.1: Indian Pompano.

potential for the marine finfish aquaculture sector. It is a pelagic species which inhabits shallow coastal waters. Indian pompano is most common in shallow coastal waters in a number of environments, including coral and rocky reefs and shore faces and tidal flats. The species has a wide salinity tolerance, as evident from the ranges from which juvenile and sub-adult fish are caught in Indian waters. The young fish eventually move to inshore reefs as they mature, before again moving to deeper outer reefs. It is distributed in western Indian Ocean from the Gulf of Oman eastward to Sri Lanka. Its range also extends to Singapore, Gulf of Thailand and Hong Kong. In India it has been reported both from the east and west coasts. Indian pompano is reported to grow to a size of 90 cm in total length and 8.1 kg body weight. Under captivity, when grown in offshore cages, Indian pompano grew from 42.8 g to 969.90 g after 9 months of culture.

Broodstock and Spawning

Broodstock Acquisition and Transport

Initially, Indian pompano broodstock can be developed through collecting or purchasing wild fish. The adult fish of more than 3 kg can be collected from the wild by hook and line and transported to the hatchery. The sexes are separate and the maturation size is around 3.0 kg, thus the same size fish needs to be collected in more number for developing broodstock. The other way is to collect the juvenile and raise it to the broodstock by rearing the fish in confined environment. Pond or cage cultured Indian pompano is another source of broodstock. Cage, pond or tank-reared fish are already accustomed to culture conditions and consequently easier to develop into suitable broodstock. However, it can take 2 years to grow juvenile Indian pompano up to broodstock size. The collected fish should be transported in covered tanks containing aerated or oxygenated water to reduce stress. Mild sedation, using approved sedatives for fish, such as 2 phenoxyethanol @ 50 ppm can be used to reduce stress and make handling the fish easier and safer. If the fish is collected from the culture system, they should not be fed for the previous 24 h to avoid deterioration of water quality.

Once the fish are transported to the hatchery, the fish are immediately to be shifted to quarantine area. It is to reduce the parasitic or bacterial infection. Generally the quarantine period varies from 3 to 4 weeks and can be carried out in small tank of 1 m^3 size to facilitate easy handling. During the quarantine period, broodstock management should focus on reducing the parasite load of the fish by giving bath treatment of formalin in freshwater at the rate of 30 mg L^{-1} for 15 min. Fish should be shifted to another tank after the treatment with fresh seawater. This treatment is need to be repeated every 4 to 5 days for atleast 7 to 8 times consequently. Once the fishes are free from the parasite, the fish can be shifted from quarantine area to broodstock tank.

Broodstock Tanks

Generally the broodstock tanks are used for culture and maturation of broodstock as well as for spawning. Due to the size of the broodstock and natural behavior of the fish, larger tank of 50 -100 m^3 are preferred for maintain the

Figure 11.2. Broodstock Tank.

broodstock. At Visakhapatnam Regional center of ICAR-CMFRI, 125 m^3 concrete reinforced tanks with water volume of 100 m^3 are used for broodstock development cum spawning tank. Generally tanks should be round, or square or rectangular with rounded corners, however round tank is preferable. Medium-range blue, green or grey colour is preferable for the broodstock tank. Generally, the broodstock tank is provided with a flow through system, however re-circulating aquaculture system is better for broodstock development and spawning. An overflow pipe is fixed 15 cm down from the surface to the egg collection tank with 500 μm mesh installed in the egg colleting chamber for sieving the eggs. The egg collecting chambers are connected with the re-circulating loop so that the water should be pumped back to the tank. It is advisable that broodstock area should be roofed in order to reduce the growth of algae on the tank wall and bottom, which makes egg collection difficult and increase the risk of failure of re-circulating system. Moreover, dirty tanks need to be cleaned frequently which may stress the broodstock and cause spawning failure or lower the quality of eggs. Broodstock tanks should have re-circulating facilities with 300 per cent water re-circulation. Sea water used for broodstock development should be filtered and clear with stable salinity of 30-35 ppt and water temperature of 27-32°C. Tanks should have natural photoperiod.

Broodstock Management

The male and female of Indian pompano cannot be identified morphologically. The sex of individual fish can be confirmed only by physical examination. Fishes need to be anaesthetized by using 2 phenoxy ethanol at the rate of 200 ppm for 2-3 minutes. The anaesthetized fish is then cannulated using the fish cannula or baby feeding tube CH 6 having an inner diameter of 1 mm and outer diameter of 2 mm. The cannula is inserted into the urinogenital orifice of males and the oviduct of females. Fish to be cannulated are anaesthetized and a wet cloth or towel is placed over the eyes in such a way that it should not cover the operculum to assist in calming the fish.

Figure 11.3. Cannulation of Indian Pompano using Baby Feeding Tube.

The cannula is guided into the fish for a distance of 6–7 cm and suction is applied to the other end of the cannula as it is withdrawn. After withdrawal, the sample within the cannula is expelled onto a microscope slide for immediate examination or into a vial containing 1 per cent neutral buffered formalin for later measurement of egg diameter. Generally, females in spawning condition will have oocytes over 500 μm diameter. To confirm matured male, the abdomen of an anaesthetized fish is gently massaged in a head-to-tail direction. A sexually ripe male spawner will extrude copious milt from its urinogenital pore. If no milt is expressed, the fish is immature, and is not ready for spawning. At VRC of CMFRI, broodstock are fed to satiation at least once daily in the morning with fresh or frozen squid and clam meat. Various vitamins namely, vitamin A (25,000 IU, USV limited, Nani Daman, India), vitamin B-complex (Pfizer, India), vitamin C (500 mg; Abbott Healthcare Pvt. Ltd., Thane, Maharashtra, India), vitamin E (400 mg) (Merck, Goa, India) and vitamin–mineral mix (Agrimin Forte, Virbac Animal Health India Pvt. Ltd., Mumbai, Maharashtra, India) were supplemented twice a week along with the feed to avoid any possible nutritional deficiencies. These vitamin pre-mix tablet are inserted inside the squid and fed one by one till the satiation of the fish.

Induced Spawning

Indian pompano needs to be stimulated with the use of hCG hormone for the spawning. Once the female ova size attained a size of more than 500 μm, and the males are oozing, then both the sex need to be injected hCG at the rate of 350IU/kg body weight. The fishes for cannulation should be anesthetized to avoid stress during the cannulation and injection. As far as possible, avoid the checking of male before inducing for spawning. Then the fishes are shifted back to the same tank for spawning. The fishes respond after 36 h of injection at a temperature range of 28-32

°C. The re-circulating system is to be stopped by 7.00 pm to avoid washing out of eggs. Generally the Indian pompano will respond in night between 11.00 to 12.00 PM.

Egg-handling Procedures

Collection

The fertilized eggs of Indian pompano are floating in nature at salinity of more than 30 ppt. This floating nature of eggs is helpful in colleting eggs by overflowing egg in collecting chamber. The fertilized eggs of Indian pompano are non-adhesive and pelagic, and range from 0.9 to 1.0 mm in diameter. A 500 µm mesh size bag is tied in egg collecting chamber, which is connected to broodstock tank via a 3 inch PVC pipe. The water level in the broodstock tanks should be raised and the re-circulating system should be started in the morning by 9.00 AM by switching on the water flow from the egg collecting chamber to the re-circulating system. The eggs collected in egg colleting chamber can be scooped with the help of 500 µm mesh size sieve.

Disinfection

The collected eggs should be disinfected with 20 ppm active iodine for 10 minutes in strong aeration. The treated eggs are then washed with deionized sea water and stocked in glass aquarium with water salinity of 30-32 ppt. As noted earlier, fertilized eggs will float on the water surface whereas unfertilized or dead eggs settle in the bottom. Only floating eggs are used for incubation and hatching because these are more likely to be good quality than sinking eggs, which are unfertilized or dead.

Qualitative Evaluation of the Eggs

Egg quality in marine finfish is generally evaluated using both qualitative and quantitative methods. Fertilized eggs are examined under a microscope (4x or 10x) for qualitative evaluation of embryos. The good fertilized eggs will have following characters:

- ☆ Eggs should be regular in shape
- ☆ During the early stages of embryonic development, the individual cells should be regular in size
- ☆ Eggs and embryos should be transparent, with no dark areas
- ☆ Chorions (egg shells) should be free of any parasites or fouling organisms.

Quantitative Evaluation of Eggs

Fertilization and hatching rates are also used as indicators of egg quality.

Fertilization Rate

Estimation of fertilization rate is undertaken several hours after fertilization has taken place, but well before hatching, as embryonic development makes it easier to discriminate the fertilized eggs (Figure 11.4) from the unfertilized eggs (Figure 11.5). For estimating fertilization rates, at least 10 egg samples are taken

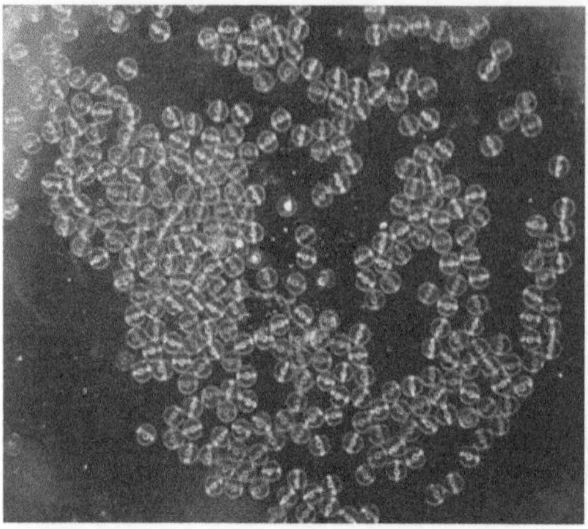

Figure 11.4. Fertilized Eggs of Indian Pompano, Embryo Development gives them a Clear View.

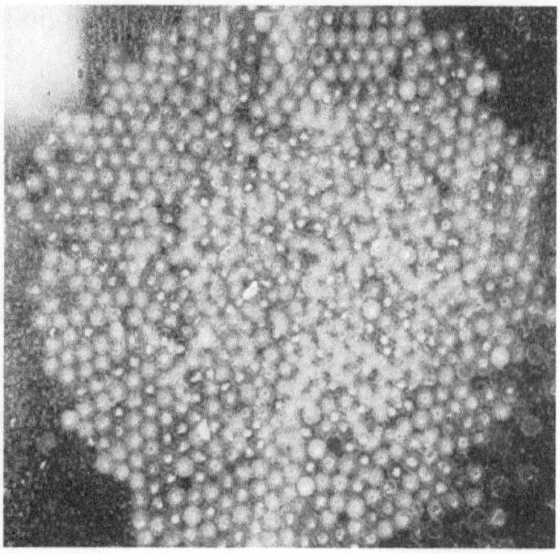

Figure 11.5. Unfertilized Eggs of Indian Pompano make them Opaque.

to have an accurate estimate. Before taking the sample, the egg should be mixed properly so that fertilized and unfertilized eggs should have proper representation. The number of fertilized eggs and the number of unfertilized eggs present in the samples are counted. Total number of eggs present in the sample is calculated by adding the numbers of fertilized eggs and unfertilized eggs of the sample. The numbers of fertilized eggs are divided by total numbers of eggs and multiplied by 100 to obtained fertilization rate (per cent). The fertilization rate is generally more than the 70 per cent at VRC of CMFRI.

Hatching Rate

The stocked fertilized eggs usually hatch out after 18-20 hrs after fertilization at a water temperature range of 28-30°C. Estimation of the hatching rate is undertaken when hatching is completed. The hatched out larvae drift on the water surface. The estimation of hatching rate is performed from random vertical water samples of hatching tank using PVC pipe from 10 different places. The volume of sampled water is measured and number of larvae in it is counted. The total number of larvae present in the hatching tank is calculated by extrapolating the sampled estimate for the whole tank. The total number of hatched out larvae are divided by the total number of fertilized eggs stocked and multiplied by 100 for calculating hatching rate (per cent). The hatching rate of Indian pompano at VRC of CMFRI is more than 80 per cent.

Stocking of Larval Rearing Tanks

Generally, Indian pompano eggs are stocked at the eyed stage (Figure 11.6) because they are more robust than the newly hatched larvae, at this stage. Newly hatched larvae (Figure 11.7) are very sensitive to physical shock or changes in water quality and moving them to the larval-rearing tanks may result in high levels of mortality. Because the hatching rate is not known before the eggs are stocked in

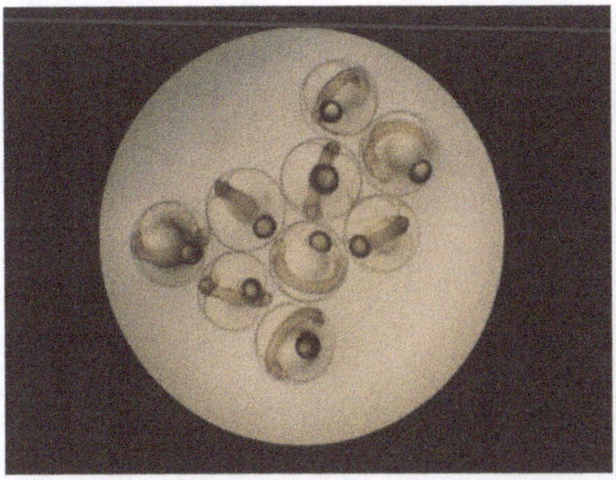

Figure 11.6. Eyed Stage Indian pompano Eggs Immediately before Hatching.

Figure 11.7. Newly Hatched Indian Pompano Larva.

the larval-rearing tanks, the number of eggs to be stocked needs to be estimated using historical hatching rates for that hatchery. Accurate estimates of the number of larvae stocked can be back-calculated using data from the actual batch stocked, as described above.

Larval-Rearing Procedures

Larval-Rearing Tanks

Generally round and rectangular tanks are used for larval rearing. At VRC of CMFRI round 2 m^3 tank with 1.2 m depth are used for larval rearing. A single aeration point in the middle of the tank was provided, which was connected with either oxygen cylinder or PSA oxygen concentrator. The aeration should be mild during the early stages (at least upto 6 days) of larval rearing to avoid physical damage of the larvae. However, it is increased gradually with progression of the larval-rearing cycle, as the larvae become more robust. The tank should be cleaned either with liquid bleach or acid wash and dried atleast for two days before stocking.

Sea water should be filtered with sand bed filter and treated with ozone at the rate of 0.1 ppm for 5 minutes. The ozonized water should be passed through carbon filter before filling in the larval rearing tank. Larval-rearing tanks should be placed under roof to avoid direct sunlight, rain and it is preferable to have larval rearing tanks inside concrete building. Light is necessary for larval rearing. It should be either diffused light or artificially provided with the help of tube lights. These tanks should be maintained at a separate quarantined area within the hatchery, with restricted entry only to authorized few persons. Every person working in the hatchery should wash their hands and feet on every entry and exit, and disinfection of all equipments should be performed before and after use.

Larval Rearing

The sea water used in larval-rearing tanks is pre-treated through sand filter to remove particulate matter and is then ozone sterilized to eliminate pathogens. The recommended initial stocking density for Indian pompano is 10 larvae/L. Live feeds used for larval rearing comprises microalgae (*Nannochloropsis* sp. and *Isochrysis* sp.), copepod nauplii and adult, small rotifers (*Brachionus rotundiformis*), large rotifers (*Brachionous plicatilis*) and brine shrimp (*Artemia*) nauplii.

The yolk sac (endogenous source) continues as the sole source of nutrients for the developing embryos immediately after hatching. The endogenous source provides nutrients for 2 days in Indian pompano larvae. Then, the exogenous feeding starts when the mouth opens after 2nd day. Their initial mouth gape is around 230 µm and hence appropriate size of feed *i.e.* copepod nauplii and screened rotifers needs to be provided. *Nannochloropsis sp* and *Isochrysis* sp in ratio of 1:3 is introduced into the larval-rearing tanks on 2nd Day of Post Hatched (DPH) at algal cell density of 1 x 10^5 cells/ml. Rotifers filtered with 100 µm mesh and copepod nauplii filtered with 100 µm mesh are introduced into the larval rearing tanks on 2nd DPH, after the larval mouth opening has been formed. The rotifer and copepod nauplii density in the larval-rearing tanks is maintained at 10-15 and 2-3 individuals/ml during 2nd-5th DPH. After 5th DPH, rotifers are introduced at densities of 20 individuals/ml, which is gradually increased to 30 individuals/ml from 8th to 10th DPH. Rotifer density gradually decreases with increase in the rate of rotifer consumption by the larvae and eventually by 13th DPH, the rotifers disappear. Freshly hatched out *Artemia* nauplii are fed at density of 0.5 individual/ml from 8th DPH and their size increasing with advancement in rearing period. Weaning of pompano larvae with artificial diets started from 11th DPH. Artificial diet with a particle size of 200-300 µm is used initially. The formulated feed is sprinkled onto the surface of the water in small amounts frequently throughout the day. Formulated feed is added in small amounts so that the feed is consumed within 5 or 10 minutes, as excess feed should not be allowed to accumulate on the bottom of the tank where it get decomposed and degrade the water quality. The size of particulate feed is increased to 400–800 µm from 22nd DPH. High-quality micro diets, specifically formulated for marine finfish, should be used and these should be stored in a refrigerator or freezer to maintain their quality.

Larval-rearing tanks are bottom siphoned on the day of hatching to remove unhatched eggs as well as hatched out eggs shells and are further needed to maintain static until 3rd DPH, and then from 4th DPH, 5-10 per cent of water exchange per day is required to maintain the rearing water quality. Bottom siphoning of the tank should be started on 4th DPH and are carried out once every 3 days. From 12th DPH, faeces, dead larvae and uneaten food accumulated on the tank bottom should be siphoned out at least once daily for maintaining water quality. Water exchange should be increased to 20 per cent per day, when both rotifers and *Artemia* are being fed together (8th DPH). Water exchange gradually increased to 50 per cent per day from 11th DPH, and is 100 per cent per day from 16th DPH.

It is of utmost importance to measure water quality regularly in the larval-rearing tanks (Table 11.1). If water quality degrades, it should be necessary to exchange the water at rates higher than the rates recommended above. However, replaced water should be of similar temperature and salinity to the water in the rearing tanks to avoid stress to the larvae.

Table 11.1. The Optimum Physico-chemical Parameters for Larval Rearing of Marine Finfish during Early Larval Rearing

Sl.No.	*Parameters*	*Value*
1.	Temperature	28-30°C
2.	Salinity	16-24 ppt
3.	Light	500 to 700 lx
4.	Photoperiod	Natural
5.	Aeration	0.62-1.25ml/min/L
6.	Dissolved oxygen	Near saturation
7.	Ammonia	<0.1 ppm
8.	Nitrite	<1.0 ppm

1st DPH

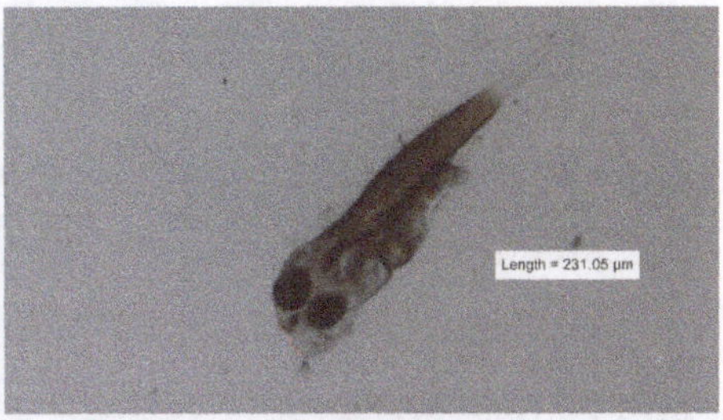

2nd DPH

3rd DPH

4th DPH

5th DPH

6th DPH

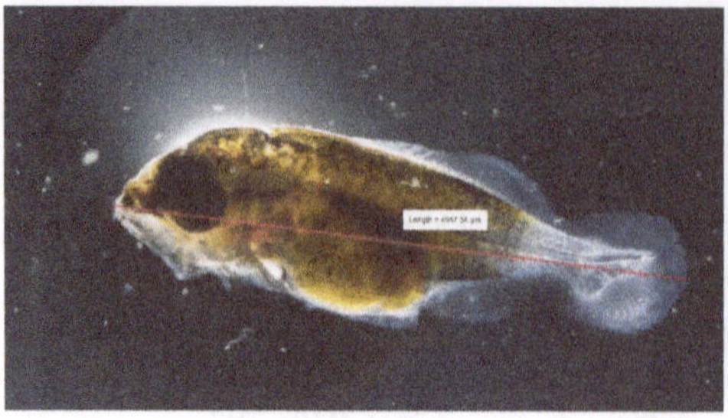

7th DPH

8th DPH

9th DPH

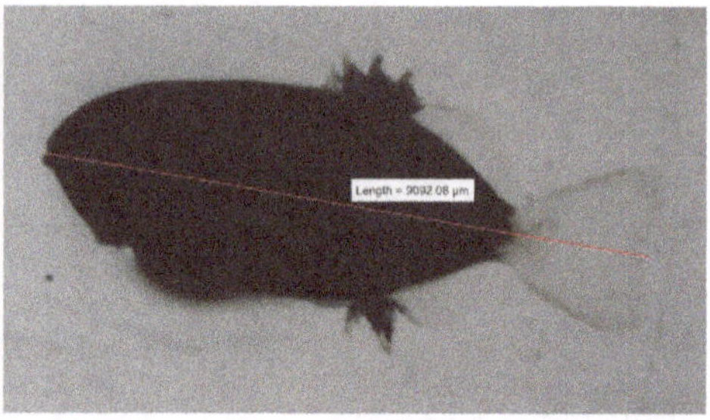

10th DPH

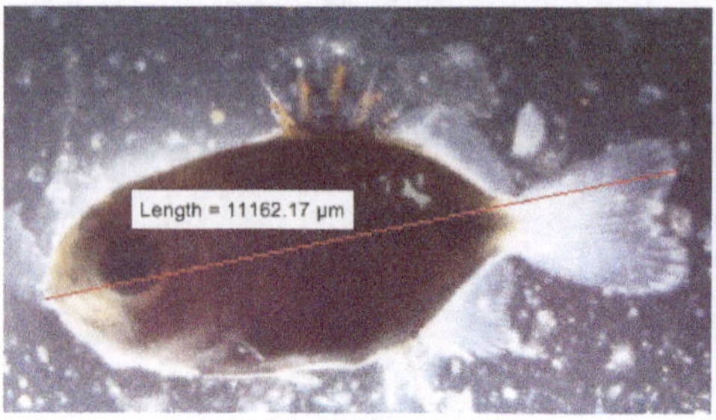

14th DPH

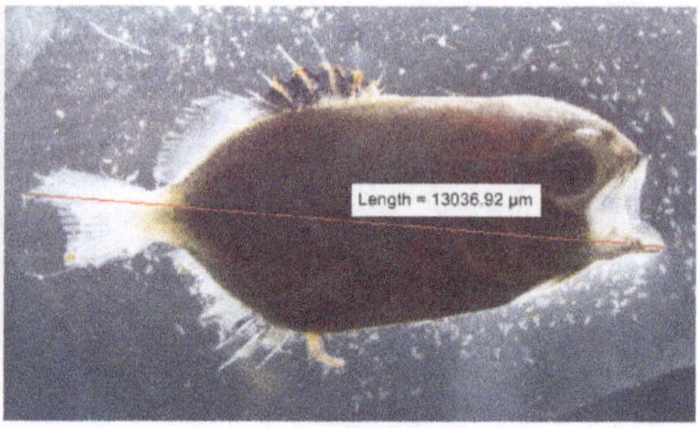

15th DPH

21st DPH

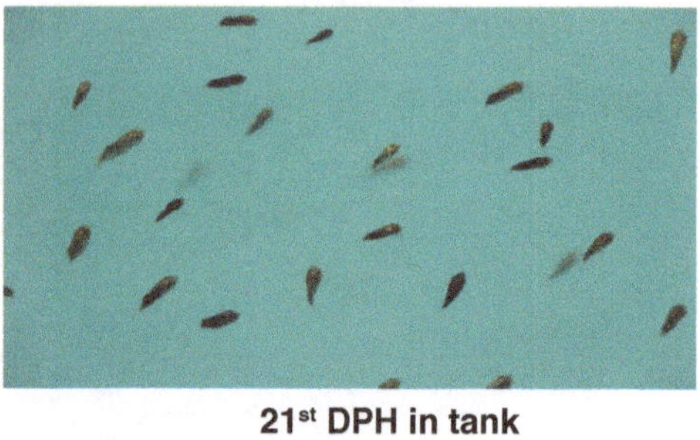

21st DPH in tank

Figure 11.8: Larval Development Stages of Indian Pompano.

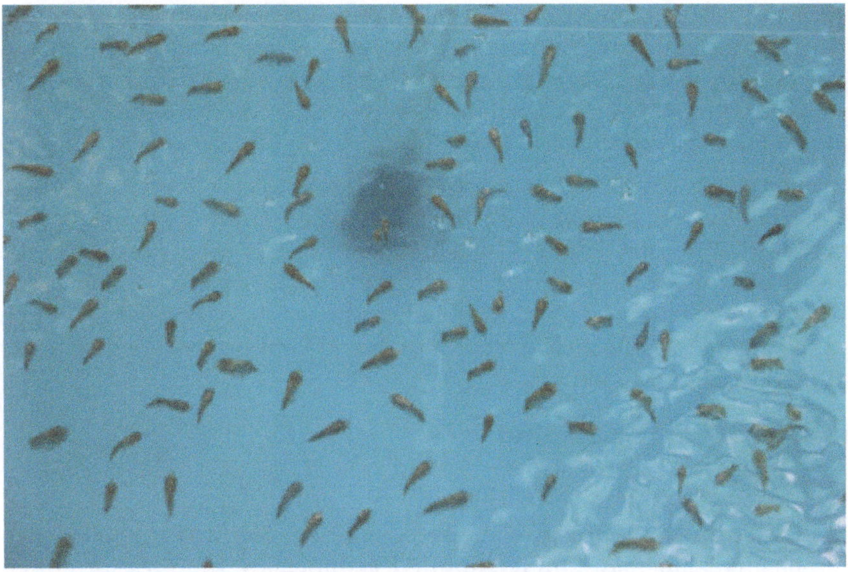

Figure 11.9. Metamorphosed Larvae of Indian Pompano.

Orange Spotted Grouper

The common name or FAO name of *Epinephelus coioides* is "orange spotted grouper". The orange spotted grouper *Epinephelus coioides* occurs in the western Indian Ocean from the southern Red Sea to Durban (South Africa) and east to the western Pacific where it is distributed from Ryukyu Islands (Japan) to New South Wales; Oceania only to Palau in the Northern Hemisphere and Fiji in the Southern; and the eastern Mediterranean. In India, this species is distributed all along the Indian coast from Gujarat to West Bengal, including the Andaman and Nicobar Islands.

Figure 11.10. Orange Spotted Grouper *Epinephelus coioides*.

Orange spotted grouper inhabits shallow reefs, lagoons, brackish water, over mud and rubble in depth to at least 30 m. Juveniles are commonly reported in the shallow waters of estuaries over sand, mud and gravel and among mangroves. They are eurythermal and euryhaline. It feeds mainly on fish followed by crabs, shrimps, squids, gastropods and bivalves.

Epinephelus coioides is a diandric protogynous species, where males are either derived from a juvenile phase or the transition of post spawning females. Females mature at 320 mm total length (TL) at an age of 2 years, whereas primary males mature at 242 mm TL at an age of 1 year. The sexual transition occurs at a TL of 550-750 mm at the age of 5-6 years. The major spawning period is March to June. They probably spawn during restricted periods and form aggregations for spawning after the full and new moon.

Broodstock and Spawning

Acquisition and Transport

Initially, orange spotted grouper broodstock are acquired through collecting or purchasing wild fish. Orange spotted grouper are diandric protogynous species where males are either derived from a juvenile phase or the transition of post spawning females. However getting males from wild is very difficult. While collecting adult fish at Visakhapatnam RC of CMFRI, Visakhapatnam, the smallest recorded size of mature male of orange spotted grouper from wild was 18 kg. Handling of such bigger size male is problematic thus collecting of smaller fishes of around 2.0 kg is good for developing male broodstock by manipulating sex using different hormones and enzymes.

Figure 11.11. Brooders *E. coioides*.

Pond or cage cultured grouper is another source of broodstock. Cage, pond or tank-reared fish are already accustomed to culture conditions and consequently easier to develop into suitable broodstock. However, it can take 2 years to grow juvenile orange spotted grouper up to broodstock size, and more over inbreeding might happen when we are collecting fishes from culture system for developing broodstock. Collected adult fish should be transported in covered tanks containing aerated or oxygenated water to reduce stress. Mild sedation, using approved sedatives for fish, such as 2 phenoxyethanol @ 50 ppm can be used to reduce stress and make handling the fish easier and safer. If the fish is collected from the culture system, they should not be fed for the previous 24 h to avoid deterioration of water quality. Generally the grouper regurgitated feed if they are handled after feeding, thus it will spoil the water quality as well as induce stress to the collected fish.

Groupers are demersal fish and hence after catching, the fish from wild shows barotraumas. Barotrauma occurs due to the differences in the pressure from where they caught and the sea surface. Due to differences in pressure, air bladder fills with air and bulge and gives a stress to the internal organs. This needs to be relieved by inserting a needle in gas bladder through anus. Once the fish are swimming normally with controlled buoyancy to maintain position in the water column, they need to be shifted in quarantine area. It is advisable to quarantine them to reduce the parasitic or bacterial infection. Generally the quarantine period varies from 1 to 2 weeks and can be carried out in small tanks of 1 m^3 capacity to facilitate easy handling. During the quarantine period, broodstock management should focus on reducing the parasite load of the fish by giving bath treatment in formalin at the rate of 200 mg L^{-1} for 30 min followed by 5 min dip in freshwater. Fish should be shifted to another tank after the treatment with fresh seawater.

Broodstock Tanks

Generally broodstock tanks are used for culture and maturation of broodstock as well as for spawning. Due to the size of the broodstock and the natural spawning behavior of the fish, larger tank of 50 -100 m^3 are preferred for broodstock tanks. At Visakhapatnam Regional Centre of ICAR-CMFRI, 125 m^3 concrete reinforced tanks with water volume of 100 m^3 are used for broodstock development cum spawning. Generally tanks should be round, or square or rectangular with rounded corners, however round tank is preferable. Medium-range blue, green or grey colour is preferable for the broodstock tank. Tanks should be at least 2.0 m deep and preferably more than 2.5 m to allow sufficient room for spawning behavior of orange spotted grouper, which involves pairs or group of fish swimming upward from the tank bottom while releasing ova and sperm. Generally the broodstock tanks are used as flow-through system, however re-circulating aquaculture system are better for broodstock development and spawning. An overflow pipe is fixed 15 cm down from the surface to the egg collection tank with 500 µm mesh installed in the egg colleting chamber for sieving the eggs. The egg collecting chambers are connected with the re-circulating loop so that the water should be pumped back to the tank. It is advisable that broodstock area should be roofed in order to reduce the growth of algae on the tank wall and bottom, which makes egg collection difficult and increases the risk of failure of re-circulating system. Moreover, dirty tanks need to be cleaned frequently which may stress the broodstock and cause spawning failure or lower the quality of eggs. Broodstock tanks should have re-circulating facilities with 300 per cent water re-circulation. Sea water used for broodstock development should be filtered and clear with stable salinity of 30-35 ppt and water temperature of 27-32°C. Tanks should have natural photoperiod.

Broodstock Management

Gender Identification and Tagging

The male and female of orange spotted grouper cannot be identified morphologically. The sex of individual fish can be confirmed only by physical examination *i.e.* live ovarian biopsy (LOB). Fishes need to be anaesthetized by

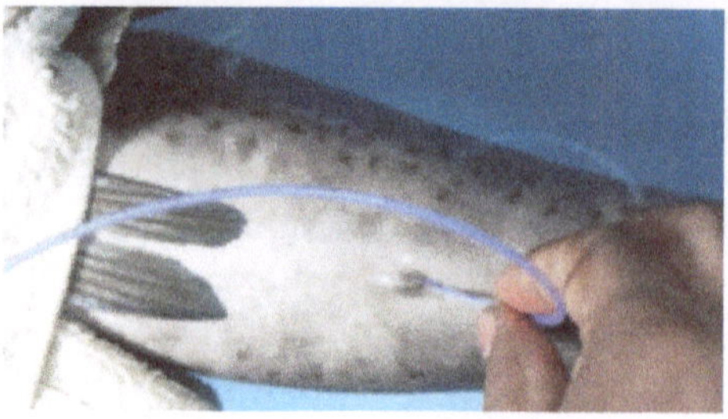

Figure 11.12. Cannulation of Organge Spotted Grouper using Baby Feeding Tube.

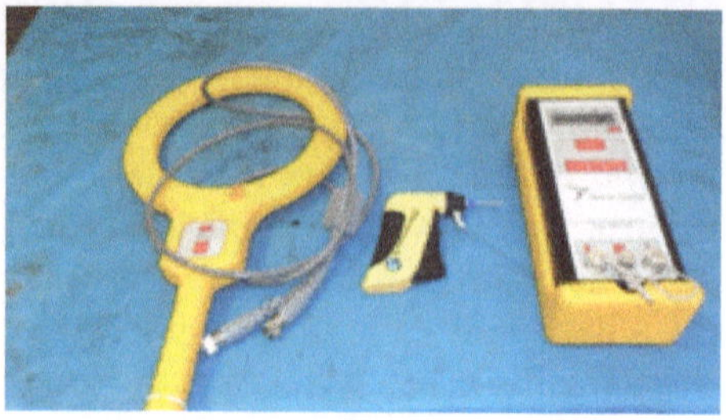

Figure 11.13. PIT Tag Used for Tagging and Identification of a Particular Brooder.

using 2 phenoxy ethanol at the rate of 200 ppm for 2-3 minutes. The anaesthetized fish is then cannulated using the fish cannula or baby feeding tube CH 6 having an inner diameter of 1 mm and outer diameter of 2 mm. The cannula is inserted into the urinogenital orifice of males and the oviduct of females (Figure 11.12). Fish to be cannulated are anaesthetised and a wet cloth or towel is placed over the eyes to assist in calming the fish. The cannula is guided into the fish for a distance of 6–7 cm and suction is applied to the other end of the cannula as it is withdrawn. After withdrawal, the sample within the cannula is expelled onto a microscope slide for immediate examination or into a vial containing 1 per cent neutral buffered formalin for later measurement of egg diameter. Generally, females in spawning condition will have oocytes over 400–500 μm diameter. To confirm matured male, the abdomen of an anaesthetised fish is gently massaged in a head-to-tail direction. A sexually ripe male spawner will extrude copious milt from its urinogenital pore. If no milt is expressed, it indicates that the fish is immature and it is not in spawning condition. Once the sexes of the fishes are confirmed, they are tagged with passive integrated transponder (PIT) and the tag number need to be maintained for future use (Figure 1.13).

Sex-reversal

As noted earlier, orange spotted grouper are diandric protogynous, and getting male from wild is very difficult thus a hatchery manager needs to manipulate the sex by using hormones and enzyme. The fifty percent of the stocked and acclimatized fishes should be implanted with pellet containing 17 α-methyl testosterone and letrazole at the rate of 5mg kg^{-1} and 0.2 mg kg^{-1} body weight respectively. The pellet can be prepared using gum acacia, cholesterol and 17 α-methyl testosterone in the ratio of 1:2:1. These chemical needs to be weighted accordingly and mixed with few drops of water, which form dough and prepared in required size as per the implanter. This prepared pellet needs to be implanted on dorsal side of the brooders below the dorsal fin in musculature. Before implantation, fish should be anaesthetized to avoid handling stress. This hormonal dose will convert the female to male within 2 months.

Feeding

At Visakhapatnam RC of CMFRI, broodstocks are fed to satiation at least once daily in the morning with fresh or frozen squid. Various vitamins namely, vitamin A (25,000 IU, USV limited, Nani Daman, India), vitamin B-complex (Pfizer, India), vitamin C (500 mg; Abbott Healthcare Pvt. Ltd., Thane, Maharashtra, India), vitamin E (400 mg) (Merck, Goa, India) and vitamin–mineral mix (Agrimin Forte, Virbac Animal Health India Pvt. Ltd., Mumbai, Maharashtra, India) were supplemented twice a week along with the feed to avoid any possible nutritional deficiencies. These vitamin pre-mix tablet are inserted inside the squid and fed one by one till the satiation of the fish.

Spawning

Orange spotted grouper broodstock are allowed to spawn naturally in the tanks. If the tank was stocked with matured female (2.0 kg) and hormone pellet implanted male, spawning will start from 3rd month onwards. Spawning generally occurs at dusk (3 pm – 6 pm) for ten to fifteen times in each month. At Visakhapatnam RC of CMFRI, grouper broodstock spawn throughout the year.

During the spawning period, orange spotted grouper may spawn between 0.8 and 6.0 million eggs each time. The re-circulating system needs to be stopped by 15.00 h during winter and 17.00 h in summer to avoid washing out of eggs. If the spawning happened then, the re-circulatory system should be in off till the next morning when the eggs are being collected. Otherwise if the spawning did not happen by 18.00 h, the re-circulating system should be on.

Stocking of Larval Rearing Tanks

Generally, orange spotted grouper eggs are stocked at later stages of eyed stage (Figure 11.14) because they are more robust than the newly hatched larvae. Newly hatched larvae (Figure 11.15) are very sensitive to physical shock or changes in water quality and moving them to the larval-rearing tanks may result in high levels of mortality. Because the hatching rate is not known before the eggs are stocked in the larval-rearing tanks, the number of eggs to be stocked is to be estimated using historical hatching rates for that hatchery.

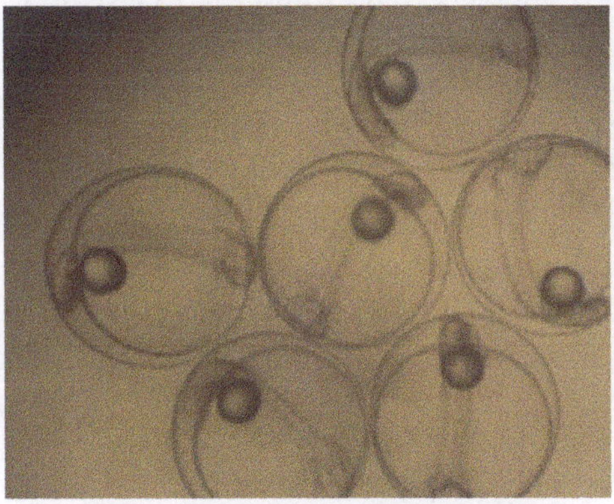

Figure 11.14. Late-Stage (Eyed Stage) Orange Spotted Grouper Eggs Immediately before Hatching.

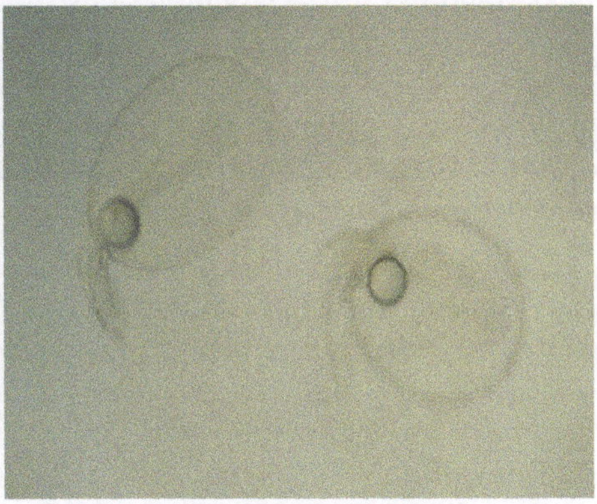

Figure 11.15. Newly Hatched Orange Spotted Grouper Larva.

Accurate estimates of the number of larvae stocked can be back-calculated using data from the actual batch stocked, as described above. If hatching rates are low, the larvae in the larval-rearing tanks should be discarded, and the tanks cleaned and disinfected.

Larval-Rearing Procedures

Larval-Rearing Tanks

Generally round and rectangular tanks are used for larval rearing. At Visakhapatnam RC of CMFRI, round 2 m^3 tank with 1.2 m depth are used for larval rearing. A single aeration point in the middle of the tank was provided, which was connected with either oxygen cylinder or PSA oxygen concentrator. The aeration should be mild during the early stages (at least upto 10 days) of larval rearing to avoid physical damage of the larvae. However, it is increased gradually with progression of the larval-rearing cycle, as the larvae become more robust. The tank should be cleaned with either liquid bleach or acid wash and dried atleast for two days before stocking.

Sea water should be filtered with sand bed filter and treated with ozone at the rate of 0.1 ppm for 5 minutes. The ozoned water should be passed through carbon filter before filling into the larval rearing tank. Larval-rearing tanks should be placed under roof to avoid direct sunlight, rain and it is preferable to have larval rearing tanks inside concrete building. Light is essential for the larval rearing and it should be either diffused light or artificially provided with the help of tube lights. These tanks should be maintained as a separate quarantined area within the hatchery restricting movement of the people inside the hatchery. The persons who enter the hatchery should wash their hands and feet on every entry and exit and all equipments before and after use should be disinfected.

Larval Rearing

The sea water used in larval-rearing tanks is pre-treated through sand filter to remove particulate matter and is then ozone sterilized to eliminate pathogens. The recommended initial stocking density for orange spotted grouper is 10 larvae/L. Oil is generally added to form a thin film on the water surface (around 0.2 ml/m^2) during 1st-4th DPH for preventing surface aggregation mortality in early-stage of grouper larvae. Live feeds like microalgae (*Nannochloropsis* sp. and *Isochrysis* sp.), copepod nauplii and adult, small rotifers (*Brachionus rotundiformis*), large rotifers (*Brachionous plicatilis*) and brine shrimp (*Artemia*) nauplii are used for larval rearing

The yolk sac (endogenous source) continues as the sole source of nutrients for the developing embryos immediately after hatching (Figure 11.16). The endogenous source provides nutrient for 2-3 days in grouper larvae. Then, the exogenous feeding starts when the mouth opens after 3rd day. Their initial mouth gape is very less so they have to be provided with appropriate size of feed *i.e.* copepod nauplii and screened rotifers. *Nannochloropsis sp* is introduced into the larval-rearing tanks on 2nd DPH at algal cell density of 1 x 10^5 cells/ml. Rotifers filtered with 80 µm mesh and copepod nauplii filtered with 100 µm mesh are introduced into the larval rearing tanks on 2nd DPH, after the larval mouth opening has been formed. The rotifer

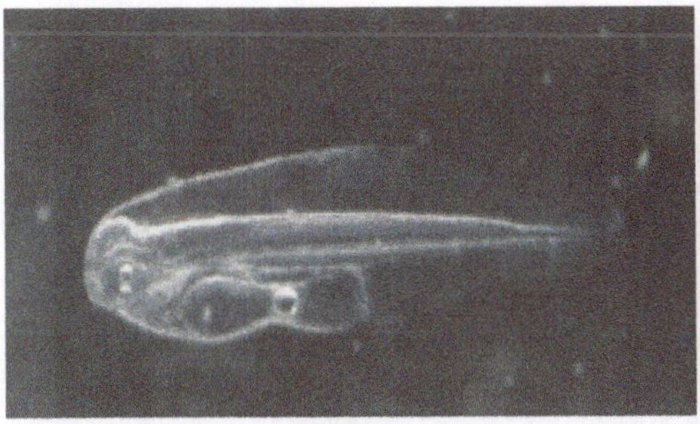

1st DPH

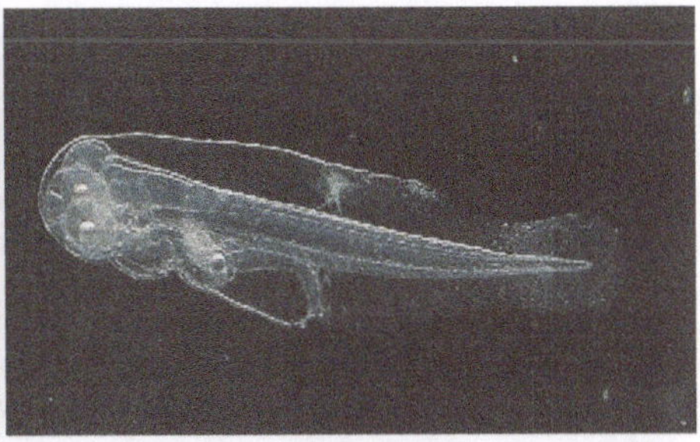

2nd DPH

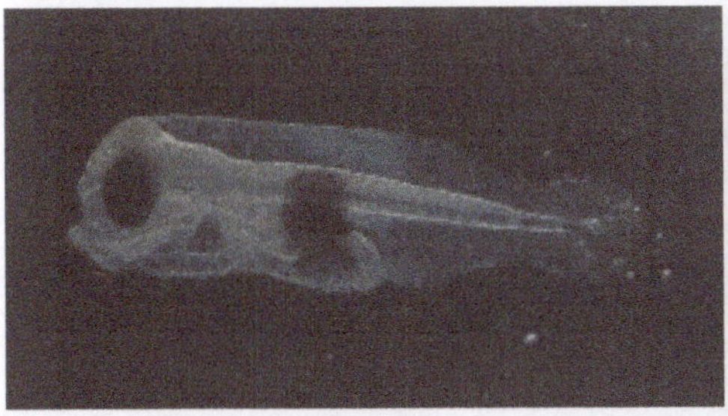

3rd DPH

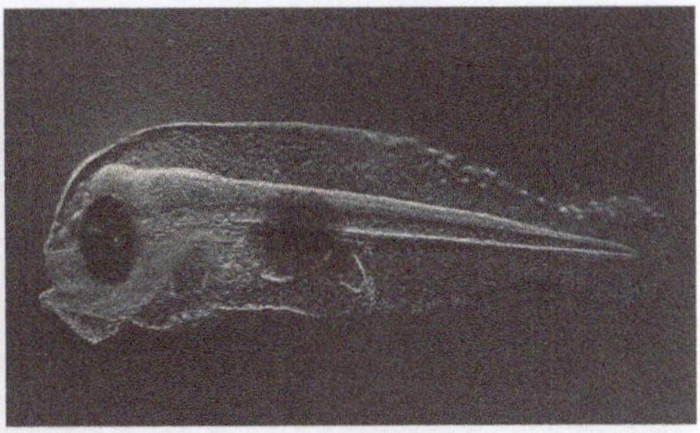

4th DPH

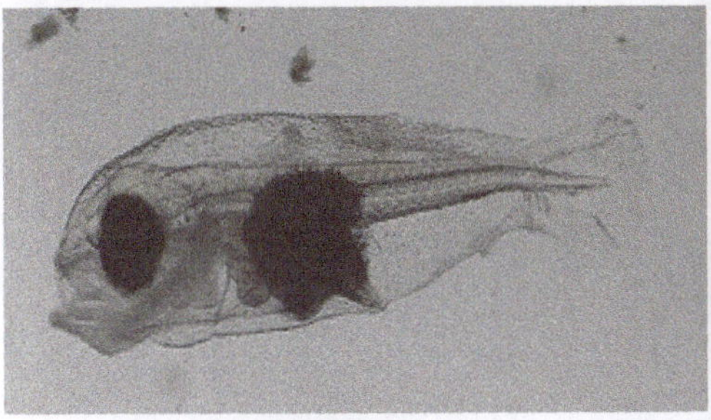

6th DPH

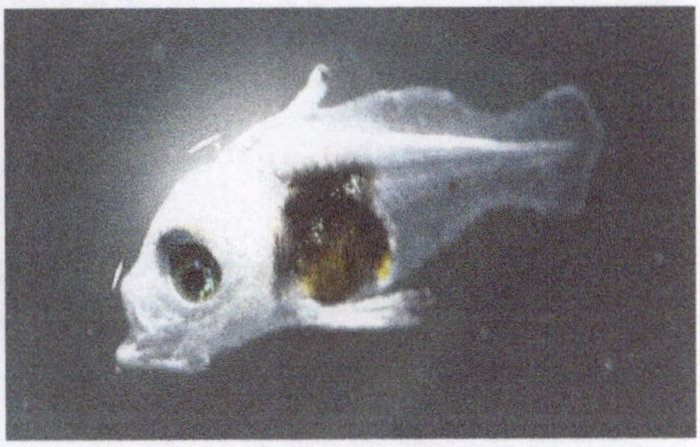

8th DPH

11th DPH

13th DPH

Figure 11.16. Larval Development Stages of Orange Spotted Grouper.

Figure 11.17. Metamorphosed Larvae of Orange Spotted Grouper (*Epinephelus coioides*).

and copepod nauplii density in the larval-rearing tanks is maintained at 5-7 and 2-3 individuals/ml respectively during 2nd-5th DPH. After 5th DPH, small rotifers (filtered with 150 µm mesh) are introduced at densities of 10-15 individuals/ml, which is gradually increased to 20 individuals/ml from 11th to 18th DPH. Rotifer density gradually decreases with increase in the rate of rotifer consumption by the larvae and eventually by 30th DPH, the rotifers disappear. Freshly hatched out *Artemia* nauplii are fed at density of 0.5 individual/ml from 17th DPH and their size increasing with advancing in rearing period. Adult copepods are fed during 16th-20th DPH in larval rearing. Weaning of grouper larvae with artificial diets starts from 20th DPH. Artificial diet with a particle size of 200-300 µm is used initially. The formulated feed is sprinkled onto the surface of the water in small amounts frequently throughout the day. Formulated feed is added in small amounts so that the feed is consumed within 5 or 10 minutes, as excess feed should not be allowed to accumulate on the bottom of the tank where it get decomposed and degrade water quality. The size of particulate feed is increased to 400–800 µm from 30th-45th DPH. High-quality micro diets, specifically formulated for marine finfish, should be used and these should be stored in a refrigerator or freezer to maintain their quality. In addition, minced fresh fish meat is fed from 30th DPH.

Larval-rearing tanks are maintained static until 7th DPH, and then from 10th DPH, 5-10 per cent of water exchange per day is required to maintain the water quality. Bottom siphoning of the tank should be started on 7th DPH. From 12th DPH, faeces, dead larvae and uneaten food accumulate on the tank bottom should be siphoned out once every day for maintaining water quality. Every day 20 per cent water exchange should be done every day when both rotifers and *Artemia* are being fed together (15th-20th DPH). Daily water exchange should be gradually increased to 50 per cent from 25th DPH onwards and 100 per cent exchange should be carried out from 35th DPH onwards.

It is very important to measure the water quality parameters regularly in the larval-rearing tanks (Table 11.2). If water quality degrades, it should be necessary to exchange the water at rates higher than the rates recommended above. However, to avoid stress to the larvae, replaced water should have equal temperature and salinity of the rearing tanks.

Table 11.2. The Optimum Physico-chemical Parameters for Larval Rearing of Orange Spotted Grouper during Early Larval Rearing

Sl.No.	*Parameters*	*Value*
1.	Temperature*	28-30°C
2.	Salinity	16-24 ppt
3.	Light	500 to 700 lx
4.	Photoperiod*	Natural
5.	aeration	0.62-1.25ml/min/L
6.	Dissolved oxygen*	Near saturation
7.	Ammonia*	<0.1 ppm
8.	Nitrite*	<1.0 ppm

* Reported for other species of grouper.

References

1. Caberoy, N. B. and Quinitio, G. F. 1998. Sensitivity of grouper *Epinephelus coioides* eggs to handling stress at different stages of embryonic development. The Israeli Journal of Aquaculture-Bamidgeh, 50: 167-173.
2. Craig, M. T., Sadovy de Mitcheson, Y. J. and Heemstra, P. C. 2011. Groupers of the world. A field and market guide. NISC (Pty) Ltd. Grahamstown, South Africa, 47pp.
3. De Mitcheson, Y. J. S. and Heemstra, P. C. 2011. Groupers of the world: a field and market guide. North America: CRC Press/Taylor and Francis Group, 356 pp.
4. Ghosh, S., Muktha, M., Behera, P. R., Megarajan, S., Ranjan, R., Gopalakrishnan, A. 2017. Validation of *Epinephelus coioides* (Hamilton, 1822) occurrence along north-east coast of India. Indian Journal of Geo Marine Sciences 46(02), 266-271.
5. Grandcourt, E. M., Al Abdessalaam, T. Z., Francis, F., Al Shamsi, A. T. and Hartmann, S. A. 2009. Reproductive biology and implications for management of the orange-spotted grouper *Epinephelus coioides* in the southern Arabian Gulf. Journal of Fish Biology, 74, 820–841.
6. Heemstra, P. C. and Randall, J. E. 1993. Groupers of the world (Family Serranidae, Subfamily Epinephelinae). An annoted andillustrated catalogue of the grouper, rockcod, hind, coral grouper and lyretail species known to date. FAO Species Catalogue, FAO Fisheries Synopsis no. 125, Rome.
7. http://www.fao.org/fishery/culturedspecies/Epinephelus_coioides/en downloaded on 07 December 2016.
8. http://www.fishbase.org/summary/6465 downloaded on 07 December 2016.
9. http://www.iucnredlist.org/documents/attach/44674.pdf downloaded on 07 December 2016.
10. Randall, J.E., 1995. Coastal Fishes of Oman. University of Hawaii Press, Honolulu, Hawaii (439 p).
11. Ranjan, R., Abdul Nazar, A.K., Jayakumar, R., 2017a. Trachinotus mookalee Cuvier, 1832. In: Ranjan, R., Muktha, M., Ghosh, S., Gopalakrishnan, A., Gopakumar, G. (Eds.), Prioritized Species for Mariculture in India. ICAR-CMFRI, Kochi, pp. 21–25.
12. Ranjan, R., Megarajan, S., Xavier, B., Dash, B., Ghosh, S., Menon, M and Edward, L.L., 2017. Conditioning, maturation and year round natural spawning of orange spotted grouper, *Epinephelus coioides* (Hamilton, 1822), in recirculating aquaculture system. Aquaculture Research 48 (12), 5864-5873.
13. Ranjan, R., Megarajan, S., Xavier, B., Ghosh, S., Santhosh, B. and Gopalakrishnan, A. 2018. Broodstock development, induced breeding and larval rearing of Indian pompano, *Trachinotus mookalee*, (Cuvier, 1832) – A new candidate species for aquaculture. Aquaculture 495, 550-557.

14. Ranjan, R., Megarajan, S., Xavier, B., Raju, S.S., Ghosh, S., Gopalakrishnan, A., 2019. Design and performance of recirculating aquaculture system for marine finfish broodstock development. Aquaculture Engineering 85, 90-97.

15. Ranjan, R., Xavier, B., Dash, B., Edward, L. L., Maheswarudu, G., and Rao, G. S. (2014). Domestication and brood stock development of the orange spotted grouper, *Epinephelus coioides* (Hamilton, 1822) in open sea cage off Visakhapatnam coast. Indian Journal of Fisheries, 61, 21– 25.

16. Ranjan, R., Xavier, B., Dash, B., Edward, L.L., Suresh, R. D. and Kumar, P.S., 2015. Efficacy of 17 α-methyl testosterone and letrozole on sex reversal of protogynous grouper, *Epinephelus tauvina* (Forskal, 1775) and spawning performance of sex-reversed male. Aquaculture research 46 (9), 2065-2072.

17. Ranjan, R., Xavier, B., Santosh, B., Megarajan, S., Ghosh, S., 2018. Copepod parasite *Lepeophtheirus kabatai* (Ho and Dojiri, 1977) infestation in orange spotted grouper, *Epinephelus coioides* (Hamilton, 1822) and its control in captivity. Indian Journal of Fisheries 65 (3), 122-125.

18. Smith-Vaniz, W.F., 1984. Carangidae, in: Fischer, W., Bianchi, G., (Eds.), FAO species identification sheets for fishery purposes. Western Indian Ocean fishing area 51. Vol. [Pag. Var.]. FAO, Rome.

19. Sugama, K., Rimmer, M. A., Ismi, S., Koesharyani, I., Suwirya, K., Giri, N. A. and Alava V. R. 2012. Hatchery management of tiger grouper (*Epinephelus fuscoguttatus*): A best-practice manual. ACIAR Monograph No. 149. Australian Centre for International Agricultural Research: Canberra. 66 pp.

20. Toledo J. D., Caberoy N. B. and Quinitio G. F. 2004. Environmental factors affecting embryonic development, hatching and survival of early stage lavae of the grouper (*Epinephelus coioides*). In: Rimmer, M. A., McBride, S. and. Williams, K. C (Eds). Advances in grouper aquaculture, ACIAR Monograph, No. 110. Australian Centre for International Agricultural Research: Canberra. Pp. 10-16.

21. Toledo J. D., Caberoy N. B. and Quinitio G. F. 2004. Environmental factors affecting embryonic development, hatching and survival of early stage lavae of the grouper (*Epinephelus coioides*). In: Rimmer, M. A., McBride, S. and. Williams, K. C (Eds). Advances in grouper aquaculture, ACIAR Monograph, No. 110. Australian Centre for International Agricultural Research: Canberra. Pp. 10-16

22. Toledo J. D., Golez, M. S. and Ohno, A. 2005. Studies on the use of copepods in the semi-intensive seed production of grouper *Epinephelus coioides*. In: Lee, C. S., Obryen, P. J and Marcus, N. H (Eds). Copepods in Aquaculture. Blackwell publishing Ltd, Oxford, UK. Pp 169-182

23. Toledo, J. D., Golez, S. N. and Ohna, A. 1999. Use of copepod nauplii during early feeding stage of grouper *Epinephelus coioides*. Fisheries Science, 65: 390-397.

Chapter 12

Culture of Marine Finfishes: Cage and Pond-Based Culture for Augmenting Fish Production

Sekar Megarajan, Ritesh Ranjan, Biji Xavier, P. Shiva, R.D. Suresh and Shubhadeep Ghosh

Visakhapatnam Regional Centre of ICAR-Central Marine Fisheries Research Institute, Visakhapatnam

Aquaculture is one of the fastest growing food-producing sectors globally, having the greatest potential to meet the ever-growing demand of aquatic food around the world. In India, the population growth is projected to be 1.65 billion by 2050 and it has been estimated that India has to produce nearly 10.5 million tonnes of marine fishes to meet the food requirement of these growing population. To achieve the huge fish production, significant contribution from mariculture/coastal aquaculture is essential in addition to the stagnated marine fish production of 3.5-4.0 million tonnes reported for last few years. India is bestowed with the vast resources for mariculture/coastal aquaculture including 6.75 million hectares of salt affected inland saline soils, but still the development of the coastal mariculture is in the preliminary stage when compared to the global scenario. Therefore, several technologies have been introduced by the research institutions for increasing the cultured fish production through mariculture.

Among all, culture of marine indigenous and exotic shrimp has become popular and mostly the coastal mariculture practices are dominated with the single group, *i.e.*, shrimp. However, dependency on single group of shellfish culture has not lead to radical increase in the fish production of the country through mariculture. Shrimp culture sector has been frequently confronted with several issues and if it alone is being continued, it may lead to un-sustainability of mariculture in the future. Therefore, species diversification is considered to be one of the major alternative methods for sustainable aquaculture development in the country. Understanding the importance, recently seed production technologies for high value marine finfishes have been developed and culture possibilities of the fishes have been studied in different culture systems including sea cages and coastal ponds and the results has been encouraging.

Orange spotted grouper *Epinephelus coioides* is highly prized for the flesh quality and fetches high market price in the world Live Reef Food Fish (LRFF) trade. It considered as one of the potential candidate species for aquaculture because of its several culture characters like fast growth, hardy in nature with tolerance to range of water salinities and preferable taste with high market value. In India, the fish fetches around Rs 300-400/kg in wholesale market, whereas in live fish international market it fetches 3-4 times higher price than the dead fish. Indian pompano (*Trachinotus mookalee*) is also considered to be a good candidate species for aquaculture due to its fast growth rate, easy adaptability to the culture condition, quick acceptance to artificial feed, euryhaline nature, pleasant appearance, good meat quality and high consumer preference.

Cage Culture of Marine Finfishes

In India, open sea cage culture technology is a new and relatively recent activity. Understanding the importance of cage culture, the Central Marine Fisheries Research Institute has initiated cage culture as a research and development activity to identify appropriate design and suitability of cages under Indian context in the year 2006-07. The first open sea cage was launched in Bay of Bengal off Visakhapatnam coast during May 2007, but the attempt was failed due to certain technical issues. After several modifications, a 15 m cage was launched in December 2007. This cage was stocked with sea bass and harvested after 6 months with 75 per cent survival. This was the first successful harvest from open sea cage in India and this success has given motivation and encouragement to the researcher and fish farmers. For ease of operation, later 6 m diameter cages were designed, installed and successfully demonstrated in several states of India with different commercially important finfishes and shell fishes. Based on the necessity to increase marine fish production in India, the design and development of cage farming for mariculture is essential. To accomplish the objectives, the following parameters need to be considered: 1) Suitable site and species for cage farming 2) Cage design parameters.

Site Selection

Site selection is the most important factor for the commercial viability of mariculture systems. A suitable place for cage culture should meet the following criteria: water temperature of 26°-30°C, depth of up to 10 to 15 m, close to the shore and away from polluted waters. Bays, straits and inland seas protected from strong winds and rough weather are ideal sites for cage culture. The sites should have adequate circulation of water, with wind and wave action within moderate limits.

Flowing waters with a slow current of 1.0 to 9.0 m per minute are considered ideal for cage operation. In addition to water quality and topographical parameters, legal aspects, access, proximity to hatcheries or fishing harbour, security, social and market parameters are to be considered.

Species Selection

The selection of the appropriate species of fish is necessary for the success of cage culture system. Species selected for culture should meet the following criteria: 1) Availability of seed and seed production technology 2) Good consumer demand and high market value 3) Hardy and tolerant to crowded conditions 4) Should be able to accept external source of food and 5) Fast growth rate. Fish species fulfilling above mentioned criteria will perform better in cages, and bring good returns to the farmers. Some of the important potential candidate species available in India for open sea cage culture are Groupers, Snappers, Sea breams, Cobia, Pompano and Asian seabass.

Cage Design

Presently, the floating HDPE (High-Density Poly Ethylene) cages are identified as most suitable in East and West coast of India. These cages have a buoyant frame or collar that support the bag; they are less limited than most other types of cages in terms of site requirements and can be made in a variety of designs which widely used in rough sea conditions. The floating HDPE cages are having following components:

Cage Frame

After conducting several experiments, CMFRI has identified the long lasting and high durable HDPE pipes as suitable material for cage frame design. It provides appropriate shape/structure to the cage. In addition, it also provides buoyancy and holds the system at a suitable level in the surface of the water. CMFRI has developed open sea cages of 6 m dia and 15 m dia for grow out fish culture and 2 m dia HDPE cages for seed rearing. Ideal size for grow out cage is 6 m due to its easy operation. Apart from that GI (galvanised Iron) based cages are used in high saline back water

Figure 12.1. Open Sea Cages.

Net Cage

The function of the net bag is to contain and protect the fish. The nets are kept stretched vertically with weights (ballast) at the bottom. Nets of varying dimensions and materials were tested for suitability of net cage in India. CMFRI has identified and used braided and twisted HDPE nets for grow out purpose. It can last for two or more seasons. Nylon net also can be used economically and due to low density, it requires high ballast to maintain its shape intact. The depth of net ranges from 2 to 4 m for nursery rearing and 5 to 6 m for grow out.

Mooring System

It holds the cage in the suitable position according to the direction and depth decided in the design, and sometimes helps to maintain the shape of the cage. The mooring joins the cage at the anchor system. The materials used in the mooring systems are steel lines, chains, reinforced plastic ropes and mechanical connectors. A mooring system must be powerful enough to resist the worst possible combination of the forces of currents, wind and waves without moving or breaking up. Presently, in India 14 mm alloy steel link chains with 'C' hook or 'U' shackles are used in cages. In mooring system, 2-3 barrels (1 m length with 220 cm dia) filled with Polyurethane Foam (PUF) is used for floatation.

Anchor System

It holds the cage and all other components at a particular site in the seabed and is connected to the cage by the mooring system. In India, dead weight anchors are commonly used. These are usually concrete blocks and the major advantage with this type is that they are fairly consistent in holding. Initially,gabion boxes were used by CMFRI for anchoring purpose; but now concrete blocks of approximately 250 kg weights are being used. The anchor system is connected to the mooring system by ropes and chains. The anchor system is formed by a system of concrete blocks of minimum 3 tonnes joined together by chains to provide appropriate strength and connected to a buoy by a braided rope

Seed Resources for Cage Culture

Open sea cage culture has been successfully demonstrated in different parts of India, and it is expected to grow in coming years. Availability of adequate number of marine finfish seeds is most important for cage culture operation. Presently, seeds are either sourced from hatchery or wild caught. In India, hatchery production technology is available for only few marine fishes like sea bass, orange spotted grouper, cobia and pompano. The commercial scale hatchery operations for sea bass and cobia produces only restricted number of seeds in a year. Cage culture has also been demonstrated for other species, but they have to depend upon on wild caught seed, since hatchery technology has not been developed in India. The other common cultivable marine candidate species in cages include sea bass, groupers, snappers and sea bream. These species are collected from the wild and stocked in cages using capture based aquaculture (CBA) technology

Demonstration of Cage Culture in India

Successful demonstration of open sea cage culture for finfish and shellfish was carried out by CMFRI in the coastal states of Gujarat, Maharashtra, Karnataka, Kerala, Tamil Nadu, Andhra Pradesh, Orissa and West Bengal. Lobster culture was successful at Kanyakumari, Vizhinjam and Mandapam. Asian seabass culture has shown highly encouraging result at Karwar, Balasore and Chennai. Mullets, Asian seabass and Pearl spot were also successfully harvested from backwater cages at Cochin and Mangalore. Cobia was successfully demonstrated and harvested in Mandapam and Karwar. Malabar grouper was successfully harvested in Palk Bay. Indian pompano and orange spotted grouper from Visakhapatanam Regional centre of CMFRI. Apart from grow-out culture, presently marine fish brood stocks are also being developed in cages with successful gonad development for Cobia and silver Pompano at Mandapam and Indian pompano and orange spotted grouper at Visakhapatnam.

Cage Culture of Fishes

Indian Pompano

Trachinotus mookalee belongs to the family Carangidae (jacks and pompanos) holds immense potential for the marine finfish aquaculture sector. It is a pelagic species inhabiting shallow coastal waters. It is distributed from the Western Indian Ocean to the Western Pacific Ocean. Indian pompano is reported to grow upto a size of 90 cm in total length (Randall, 1995) and 8.1 kg body weight (Smith-Vaniz, 1984). It is a promising candidate species for mariculture due to its attractive appearance, fast and uniform growth rate, adaptability to culture environment, acceptability to formulated feed, firm, white as well as tasty meat and high market demand. Indian pompano is caught sporadically in the commercial fishery and therefore its availability is rather scarce in the fish markets and hence the increasing demand can only be met through aquaculture.

Nursery Rearing of Indian Pompano

Nursery rearing of Indian pompano has been standardized with different feed and culture conditions. In rearing tanks, when nursing the fry at a density of 150 Nos./m^3, having average weightof 3.95 g reaches 28.08 g size in a period of two months. Fry are fed with artificial pelleted feedcontaining 45 per cent protein and 10 per cent fat @ 10 per cent of biomassfor four times a day. In ponds, Indian pompano fry weighing averagely with 2 g are nursed in hapas at a density of 150 Nos./m^3 and the fry attains a weight of 20 g after 60 days of rearing and the feeding regime followed is similar as mentioned earlier. In marine cages (HDPE of 6 m diameter with 8 mm size inner net), hatchery produced seeds weighing 2.5 g and 5.25 cm are stocked at 35 Nos./m^3 and are fed at 10 per cent of body weight with commercial floating diet containing 45 per cent crude protein and 10 per cent fat twice a day. After three months of culture, fishes attain an average body weight of 50 g.

Grow Out Culture of Indian pompano

Indian pompano fishes with an average body weight of 50 g are stocked in

HDPE circular cages of 6 m diameter (inner net of 8 mm mesh) at 35 Nos./m^3 and are fed with floating pellets having 40 per cent crude protein and 10 per cent fat. The fishes are reared for four months till they reach an average body weight of 280 g and thereafter, fishes are stocked into two different stocking densities *i.e.* 15 and 20 Nos./m^3 in two different 6 m dia HDPE cages with inner net of 2.5 cm mesh size. Henceforth, fishes are fed with low value finfishes (sardine, Indian scad and tilapia, *etc.*) at 8-10 per cent of body weight twice a day for a culture period of another five months. During the initial four months of rearing, fishes grow from 50 g to 280 g with an average FCR of 1.0:1.29 on artificial pelleted feed. In the next five months of culture, fishes stocked at 15 Nos./m^3 reach to 769 g with a calculated FCR of 1.0:4.98 and fishes stocked at 20 Nos./m^3 reach to 478 g with a calculated FCR of 1.0:7.48. Specific growth rate for the fishes stocked at 15 Nos./m^3 and 20 Nos./m^3 are 1.83 per cent and 1.69 per cent per day respectively with an average survival of 93.6 per cent.

Figure 12.2. Indian Pompano in Sea Cage.

It is observed that feed acceptance of the fish for pelleted feed and low valued fishes are equally good. However, among low valued fishes, acceptance for sardine and Indian scad is comparatively better than tilapia. The expected production is 1.975 Tons and the fishes can be sold at a farm gate price of Rs. 350/kg.

Orange Spotted Grouper

Epinephelus coioides is an ideal candidate species for aquaculture in tropical and subtropical waters. It belongs to the family Serranidae and sub-family Epinephelinae. It is commercially important fish, particularly for live seafood markets in several Asian countries such as Hong Kong, China, Taiwan, Singapore and Malaysia. It is highly prized for the quality of flesh and fetches high market prices (Johnston and Yeeting, 2006). *Epinephelus coioides* is a diandricprotogynous species, where males are either derived from a juvenile phase or the transition of post spawning females. Females mature at a total length (TL) of 320 mm in the age of 2 years, whereas primary males mature at 242 mm TL at an age of 1 year. The sexual transition occurs at a TL of 550-750 mm at the age of 5-6 years. The major spawning period is March to June. They probably spawn during restricted periods and form aggregations for spawning after the full and new moon. Majority live on coral reefs, but are also found in estuaries, on rocky bottoms and less frequently in sandy or silty habitats (Heemstra and Randall, 1993).

Nursery Rearing of Orange Spotted Grouper

The nursery rearing of orange spotted grouper has been standardized with different feed and culture conditions. Nursery rearing of grouper comprises of two phases. In the first phase, 2.5-3.0 cm (0.4 g) fry are cultured in a tank for period of two weeks till they accept artificial feeds completely, by which time they reach upto 5-6 cm. The fry during this period is reared in 1 Ton capacity tank @ 1No./l. They are fed on artificial diet and *Artemia* biomass for completely weaning the larvae to artificial feed. Artificial feed containing 45 per cent protein and 10 per cent fat are used in this period. Grading is performed every 5 days to grade the larvae according to size. The second phase consists of growing 5-6 cm size to 10-15 cm for stocking in cage as well as in pond. During this period, the fingerlings are reared either in pond based hapa (2 mm mesh size), flow through cement tank or in re-circulatory system. Pellet feed with different protein level and minced fish meat are used as feed, among which pellet feed with 45 per cent protein yields higher growth. The rearing system is also found to influence the growth rate, where highest average daily weight gain of 0.59 g/day is observed in RAS, followed by 0.4 g/day in pond and 0.26 g/day in cement tanks after one month of rearing, when stocking fingerlings of 2-3 g and feeding with 45 per cent pellet feed @ 10 per cent body weight daily. In all the rearing systems, fingerlings of 2-3 g stocked grow to an average size of 15-20 g after a month when fed with pellets containing 45 per cent protein.

Figure 12.3. Cage Culture of Orange Spotted Grouper.

Grow Out Culture of Orange Spotted Grouper

HDPE cages (6 m in diameter and 4 m net depth) are used for the grow-out culture of orange spotted grouper. Hatchery produced fingerlings of orange spotted grouper (average length of 10 cm and average weight of 15 g) are stocked at two different densities, 10 Nos./m^3 and 20 Nos./m^3 in cages. The fingerlings are fed with pelleted feed containing 40 per cent protein and 10 per cent fat @ 10-8 per cent of body weight, twice a day for the first three months of culture. After three months, the fishes reach an average size of 20.2 cm in length and average weight of

109.4 g. The average FCR for the pelleted feed is 1:1.44. From the fourth month of culture, fishes are fed low value fish (sardines and scads) @ 15-8 per cent of body weight twice a day. After eleven months of culture, fishes reach the average size of 43.9 cm and weight of 1031 g, with average survival of 93.34 per cent and FCR of 1:6.20 when fed with low value fishes. There is no significant differences in the growth, survival and FCR of fishes when reared at 10 Nos./m^3 and 20 Nos./m^3.

Figure 12.3a. Harvesting of Cage Cultured Orange Spotted Grouper.

Initially the fishes are stocked in cage with 1 cm mesh size, after reaching 100 g size, fishes are transferred to 2 cm mesh size nets and finally on reaching 500 g size, they are stocked at 5-6 cm meshes till harvest. Periodical sampling, net cleaning and mooring checking are to be done at regular fortnight intervals and net is exchanged after every three months of culture. When stocked with 10 Nos./m^3 and 20 Nos./m^3 in the cages, the average production is 1.12 T and 2.14 T respectively. Partial harvesting once in a week is done and the fishes are sold at a farmgate price of Rs. 350/kg.

Pond Culture (Case Study)

Culture of Indian pompano and Orange spotted grouper can be performed in ponds. Usually, abandoned shrimp cultured ponds and newly constructed ponds are used. To culture the marine finfishes, it is recommended to use pond having depth of 1.5 m with aeration facilities. Pond preparation methods are similar to pond preparation for culture of shrimps.

Pond Preparation and Water Treatment

The existing shrimp farm can be selected for fish culture. Pond preparation is started with pond drying and treating the soil with lime (quick lime @ 250 kg/ha) and then filled with creek water up to a depth of 1.5 m. All animalcules from the pond are to be killed by chlorination of the pond at 10 ppm and then removed from the pond. After four days, the chlorinated water is fertilized by application of urea (2.5 ppm) and TSP (3 ppm) to condition the pond water quality.

Pond Culture of Indian Pompano

Stocking and Nursery Management

Hatchery produced advanced Indian pompano fry seed with the size of 0.5 to 1 g size, were transported to farm site located at Bhavedevarapalli. The seeds were packed at 50 Nos./l in polythene bags filled with oxygen and transported in low temperature. Upon reaching to the site, the seeds were initially stocked into the hapa in culture pond. Seeds were stocked in a hapa having mesh size 0.8 to 1.5 cm with dimension of 2x2x1.5 m at the stocking density 250-300 Nos./m^3. A total of 3250 numbers of seeds were stocked in a single pond with 1 acre area.

The stocked fish fingerlings were maintained in hapa with aeration and fed 4 times in a day. Initially seeds were fed @ 8-10 per cent of bodyweight and then slowly reduced to 5 per cent after reaching 25-30 g. During the nursery period survival rate of >90 per cent was observed. During the nursery rearing, the fish fingerlings were monitored for their growth and health for every fortnight. Fish fingerlings were fed with commercially available pellet feed from 1) Growell feeds Pvt Ltd (India)and 2) Uni-president Feeds (Vietnam).

Table 12.1. Feed Specifications

Sl.No.	*Feed Size*	*Nutritional Content*	*Suitable Fish Size*
1	1.2 mm	45 per cent Crude protein and 10 per cent fat – Floating feed	Up to 5-10g
2	1.8 mm	45 per cent Crude protein and 10 per cent fat – Floating feed	Up to10-50g

Grow-out Culture

The fish fingerlings were directly released in to the pond after reaching 40-50 g in size after three months period. A total of three ponds were stocked, among which two ponds were stocked @ 3000 Nos/acre directly in pond and the third pond was stocked with total of 4000 fishes, @1000 numbers/hapa erected in the pond. In cage in pond, initially fishes were stocked in 4x4x1 m cages @40 Nos./m^3 till the fish reaches 250 g and further in 8x8x1 m cages @15 Nos./m^3. The paddle wheel aerator was provided to maintain the dissolved oxygen level at above 4 ppm in the complete culture period. Fishes in pond and cage were fed four to three times in a day with commercial high protein pelleted feed. Growth parameters and health status of the fishes were periodically monitored once in fortnight. Fishes were sampled using cast net at the time of sampling. Details of feeding and growth are given below.

Table 12.2. Feed Specifications

Sl.No.	Feed Size	Nutritional Content	Suitable Fish Size	Feeding Rate
1	3 mm	40 per cent Crude protein and 10 per cent fat – Floating feed	40 – 150 g	4-5 per cent of body weight (BW)
2	4 mm		150-300 g	3-4 per cent of BW
3	5 mm		300-500 g	2-3 per cent of BW
4	6 mm		500-1000g	2-1.5 per cent of BW

Figure 12.4. Grow-out Culture of Indian Pompana in Coastal Pond.

Pond Culture of Orange Spotted Grouper

Stocking and Nursery Management

Orange spotted grouper fish seed produced at Visakhapatnam Regional Centre of CMFRI were reared till 1 to 2 g size, and this advanced fry stage were transported to pond site at Bhavedevarapalli. The seeds were packed at 40 Nos./l in polythene bags filled with oxygen and transported in low temperature. Upon reaching the site, the seeds were initially stocked @ 300-400 numbers/m^3 into a hapa erected in culture pond having mesh size of 0.8 to 1.5 cm with size of 2x2x1.5 m. A total of 3000 numbers of seeds were stocked in a single pond having area of 1 acre.

Table 12.3. Feed Specifications

Sl.No.	Feed Size	Nutritional Content	Suitable Fish Size
1	1.2 mm	45 per cent Crude protein and 10 per cent fat – Floating feed	Up to 5-10 g
2	1.8 mm	45 per cent Crude protein and 10 per cent fat – Floating feed	Up to10-35 g

The stocked fish fingerlings were maintained in hapa with aeration and fed 4 times in a day. Initially seeds were fed @ 8-10 per cent of bodyweight and then slowly reduced to 5 per cent after reaching 25-30 g. During the nursery period the average survival rate of 70 per cent was observed. During the nursery rearing, the fish fingerlings were monitored for their growth and health for every fortnight. Fish fingerlings were fed with commercially available pellet feed from 1) Growell feeds Pvt Ltd (India) and 2) Uni-president Feeds (Vietnam).

Grow-out Culture

The fish fingerlings were directly released into the pond after reaching 40-50 g size reared in three-month periods. Due to the less survival during the nursery rearing stage, the fish fingerlings stocked into two ponds @ 3000 Nos./ Acre. Seed were released into the pond and fed with pelleted feed of 3 mm in size (Growell feeds Pvt Ltd). Due to the demersal behaviour, the fish did not respond properly for feeding, and hence tilapia brooders were released into the culture ponds @ 2000 Nos./Acre. The added tilapia brooders were spawned frequently and then the tilapia fry were auto-fed by the grouper.

Figure 12.5. Nursery Culture of Orange Spotted Grouper in Coastal Ponds.

Figure 12.6. Grow-out Culture of Orange Spotted Grouper in Coastal Ponds.

Simultaneously, the orange spotted grouper also were co-fed by the pellet feed at 0.5 per cent of body weight of the fish. After reaching 400 g size, the fish were also fed by the dead tilapia @ 3-4 per cent of body weight, which showed good acceptance. Growth were monitored at every fortnight, and the orange spotted grouper cultured in two ponds have exhibited huge size variation of 30 to 40 per cent differences in the total cultured fishes. 30-40 per cent of the fishes in the both ponds have reached to the average size of 1.5 kg for recording growth data, the average sizes were considered and the extreme weights were not considered for sampling.

Constraints in Mariculture Development

Several constrains are currently limiting the expansion and development of mariculture in India. The major constrains are; 1) Availability of adequate quantities of seeds and nutritionally balanced, low cost artificial feed for marine finfishes 2) Vandalism and poaching: Caged fish are an easy target for poachers and vandals 3) Market constraint, such as fluctuation of prices based on supply and quality control problems, 5) Non premium price for the cultured high value marine finfish 4) Legal and administration issues. At present in India, legal frame work for cage culture has not been developed well, regarding the controlling authority for giving legal permission for cage culture operation 5) Social issue: In future, if the cage farming will be developed on a large scale, it may have conflict with the traditional fishing community regarding accessibility to fishing grounds.

References

1. Heemstra, P. C. and Randall, J. E. 1993.Groupers of the world (Family Serranidae, Subfamily Epinephelinae). An annoted and illustrated catalogue of the grouper, rockcod, hind, coral grouper and lyretail species known to date. FAO Species Catalogue, FAO Fisheries Synopsis no. 125, Rome.
2. Johnston, B. and Yeeting B. 2006. Economics and marketing of the live reef fish trade in Asia-Pacific. ACIAR Working Paper No. 60. Australian Centre for International Agricultural Research: Canberra.
3. Randall, J. E. 1995. Coastal Fishes of Oman.University of Hawaii Press, Honolulu, Hawaii (439 p).
4. Smith-Vaniz, W. F. 1984. Carangidae, in: Fischer, W., Bianchi, G., (Eds.), FAO species identification sheets for fishery purposes. Western Indian Ocean fishing area 51.Vol. 1. FAO, Rome.

Chapter 13

Design and Construction of Biofloc Production Systems

M. Menaga, S. Felix and A. Gopalakannan

Advanced Research Farm Facility, Tamil Nadu Dr J Jayalalitha Fisheries University Madhavaram, Tamil Nadu

In the process of augmenting the production and making more profit in aquaculture, advance systems like Recirculating Aquaculture System (RAS), Raceway systems, Zero water exchange systems, Green water technology Organic farming and Aerobic Microbial Floc (AMF) technologies have been introduced in this sector to enhance fish production. Introduction of these kind of systems would bring the sector to a greater reach in the country. Of which one of the most important and highly popular system is Biofloc Technology (BFT) which yields higher production in controlled environment without much water exchange. BFT can be applied for both indoor and outdoor systems

Biofloc Pond

In outdoor system, newly constructed ponds or aged earthen ponds can be chosen for setting up of biofloc systems. These earthen ponds can be converted into biofloc pond by lining them on the side and bottom with help of low or high density polyethylene sheets. In both the kind of ponds, certain basic aspects in pond construction and design are to be considered. Mostly rectangular type of pond is preferred because of its easy maintenance, harvesting, construction cost, utilization of large water body,etc.

Once a high potential site is thoroughly analyzed and found suitable for fish farming or shrimp farming, it must be surveyed. Based on this detailed survey and the targeted production strategy, farm design plans are then drawn up by an experienced aquaculture engineer or the project manager. The possible ways to design a farm are endless, but certain designs are definitely more efficient and effective than others. Farm design is always an exciting period, but it takes a skilled and experienced aquaculture engineer to put together the best shrimp or fish farm design for a given site. Though it looks easy, it is highly complicated process and requires dedicated involvement.

The best layout would be one that gives maximum cultivable area and minimizes the area occupied by bunds and channels, which in turn reduce the cost of construction per hectare of water area of the farm. Layout planning also deals with design of all the farm elements like embankments, ponds, inlets, outlets, water channel, drainage unit,other sheds *etc.* Layout planning varies considerably from site to site, depending upon topography, level of the site, location of water resources and purpose of the farm (Helfrich and Pardue, 2009).

Type of Pond-Based on Construction

Excavated Pond

An excavated pond is often built on level terrain and its depth is achieved solely by excavation. An excavated pond is relatively safe from flood damage and can be built to expose a minimum water surface area in relation to volume. This is beneficial in areas of high evaporation losses and a limited amount of water supply. An excavated pond can be supplied by some water source such as surface runoff, water diverted from a stream or a river, water pumped from a well, surficial aquifer sources, shallow water table seepage, or any combination of the above sources.

Embankment Pond

This type of pond is built by creating an embankment or dam used to impound water and usually constructed in a valley or on gently sloping land. It is not recommended to build an embankment pond on greater than a 4 per cent slope. Less excavation may be needed to build this type of pond. The side slopes are described by using a ratio of horizontal to vertical distance along the slope. For Example, a slope of 2:1 is 2 feet horizontal to every 1-foot vertical. The slope of the side of the dam facing the water should be 2:1. The slope of the backside of the dam should be a minimum of 3:1 for stability.

Size of the Pond

The size of ponds built today can vary anywhere from 20 m^2 to 20 hectares (44 acres) or more. Pond size is determined by the type of species cultured, the intensity of the system, size and maturity of the species being farmed, access to capital, land availability, water availability, harvesting method, and even the marketing and sales goals of the project.

The species being farmed and the size of animal play a big role in pond size and farm design. For example, a commercially oriented tilapia farming operation typically utilizes 0.1 or 0.2 hectare ponds for nursery phases and 0.3 to 0.5 hectare

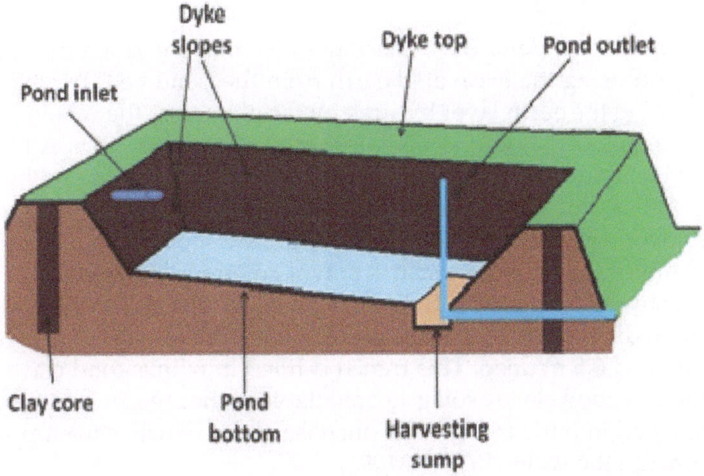

A cross section of a typical earthen fish pond showing the pond profile and important features

Dyke slope: 1:1
Bottom slope: 1% (0.01)
Free board 30cm

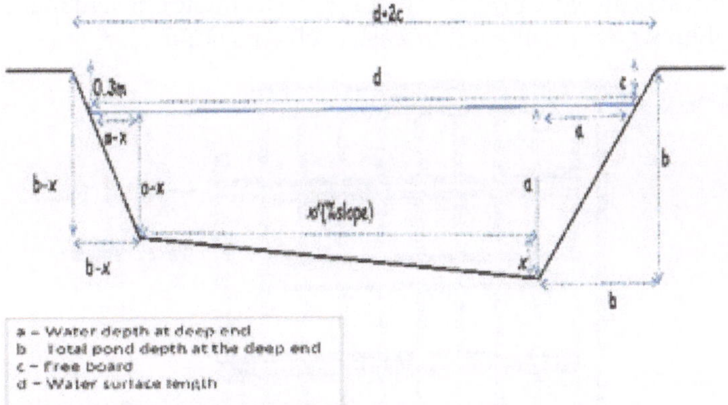

Length wise section of a typical 1:1 fishpond showing the relationship between various measurements.

Figure 13.1. Embankment Pond.

ponds for grow out. Semi-intensive shrimp farms generally use 7 to 20 hectare ponds, while more intensive shrimp farms generally use ponds less than 7 hectares and quite often less than 1-2 hectares in size. Most ponds are rectangular in shape, but there are also square, circular, and irregularly shaped ponds in existence. The excavation of ponds may be done either engaging manual labor, when speed of construction is under consideration then bulldozer or scoop may be utilized. The excavation of the pond should be as far as possible preceded in layers. The top ground soil should be stocked in one place and it should not be used for any construction.

Construction of Dykes

The construction of pond dykes becomes essential if earthen dykes are made around the pond using the excavated earth from the pond bed. All dykes should be raised dumping the earth layer by stretching right across the whole section and in such cases each layer should not be more than 30 cm thickness. All large clods should be broken and each layer should be thoroughly consolidated by watering and ramming. The sides and top of the dykes should be properly dressed and finished with wooden thappies. When the dyke is to be constructed on a sandy, gravelly or marshy soil base the construction of key trench is essential and in such cases digging should be done until water tight foundations are reached. The key trench is a small ditch or furrow dug along the line of the center of the walls about 0.5 to 1m wide and 0.5 m deep. This trench is filled in with a good clayey soil and is well rammed. If good clayey soil is not available in the area, ordinary soil should be well compacted into the trench. The purpose of the trench is to stop seepage of water under heath the walls (Woods, 1996).

Water Inlet and Outlet Structures

Both inlet and outlet should be screened to avert the entry of trash or unwanted fish and the loss of stocked fish respectively. Ponds should be constructed in a such way that they can be drained individually, completely and rapidly. This will enable the removal of all fish during harvesting and facilitate efficient management, particularly when water quality and disease problems occur.

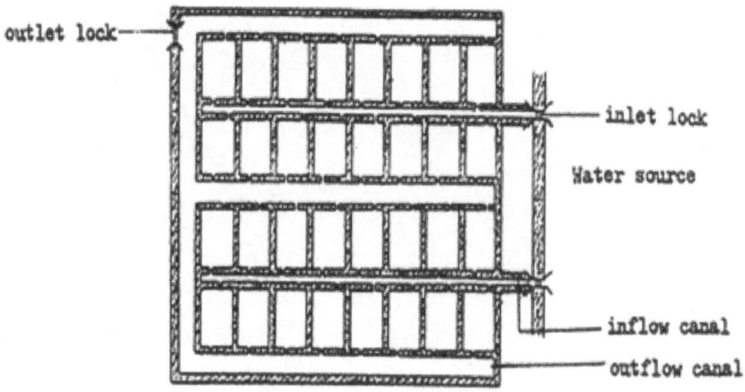

Figure 13.2. Schematic Representation of the Feeder Canals and its Types.

Different types of open canals are used to transport water to fish farms. They are divided into four types in fish farms depending upon their purpose or use.

Feeder canals: These are the canals used to supply water from the main water intake to the fish ponds. In a large farm with several diversion pond units, there is usually a main feeder canal branching into secondary and even tertiary feeder canals.

Diversion canals: These canals are used to divert the required amount of water to fish ponds or to divert excess amount of water from (ponds) source.

Drainage canals: These canals are used to remove the unwanted or polluted water from fish ponds.

Protection canals: To divert runoff water away from fish ponds or channels which carry the water to ponds.

Lining of Ponds and Dyke

The protection of slopes of the dyke against soil erosion is called as lining of the dykes. The following are the some of the important methods of lining

- ✰ Stone pitching
- ✰ Brick pitching
- ✰ Cement concrete lining
- ✰ Soil-cement lining
- ✰ Stone slab lining
- ✰ Polyethylene paper lining

In this, brick lining and HDPE lining were more advisable for biofloc ponds.

Brick Lining

It reduces the seepage at about 80 per cent provided bricks are manufactured of best material and observed less water. Bricks may be laid flat or on edge and are embedded on 1.2 cm layer of 1:6 cement mortar on consolidated pond surface. The bricks should be plastered or pointed with cement mortar of 1:3 proportions.

Figure 13.3. Lined Pond Constructed Using 330 GSM SFF-HDPE Lining Materials.

Plastic Lining

The plastic membrane or HDPE (250-700 GSM) lining offers a compact water barrier. This system offers considerable economy. In this method of lining the pond looks very simple. The pond surface should be cleared first of all from roots, vegetation *etc.* and the surface is made smooth over where the HDPE membrane is laid. Anchorages are made in the ground for holding the membrane in proper position. The membrane is then covered with 30 cm of thick earth layer and well compact it.

Figure 13.4. Aspirator (2HP aerator) Installed in Lined Pond.

Raceway and its Components

Raceway System

Raceway Type-1

To construct a raceway of 15 X 3 m dimension, required earth is to be excavated and dikes are farmed for the raceways and after sufficient consolidation, fabricated pre-cast cement slabs (50 mm thickness) are fixed on the bottom and sides of the raceway pond. The elongated raceway (15 m long; 3 m width) with rounded corners are to be lined with 700 GSM nylon fabric sheet; water outlet drain provision is given to tanks towards the end of raceways. Bottom of the raceway tanks are provided with adequate slope (0.5 per cent) towards the drain to allow easy draining of water. The outlet of raceway tanks leads to harvesting tank (3 x 1.5 x 2.1 m) and from their used water is drained to constructed wetlands (CWs). The depth of the raceway is provided to1.0 m. Further, free board of 15 cm has to be provided for the raceways for producing a maximum, of 2-3 g size shrimps in raceway tanks. Each raceway is provided with a 13 m long and 1m deep central partition. Central partition is made up of waterproof marine plywood of 20 mm thickness and fixed on cross-wooden beams with stainless steel bolts and screw. The height and position adjustment of central partition are also provided.

Raceway Type-2

Each raceways system has a capacity of 56.25 tons (15 m long; 3 m width; 1.25 m depth) with rounded corners, required earth is excavated and cement tank is constructed. Water outlet drain provision is given to tanks towards the end of raceways. Bottom of the raceway tanks are provided with an adequate slope (0.5 per cent) towards the drain pipe. The outlet of raceway tanks leads to harvesting tank (15 x 2.1m) for draining the used water. The depth of the raceway is kept as 125 cm and freeboard of 25 cm is provided. Each raceway is provided with a 13 m long and 1.25 m deep central partition to facilitate circular motion of water (Felix, 2006).

Greenhouse Structure for Raceways

The green house is constructed using cast iron pipe pillars, farming with gentle semicircle roofing of iron pipes coated with non-corrosive paint. The dimension of Green house was 20 x 10 m. The roofing is provided with 20 per cent of transparent FRP sheets and 80 per cent of non-transparent white FRP sheets. The side-walls are made up of removable green shading (75 per cent) net material fixed to wooden frames. The structure is large enough to maintain two 30 m^2 raceway with two sand filters. The green house space is equipped with 20 tube light fittings arranged in four rows. Building raceways facilities under greenhouses would help maintaining the ambient water temperature and other water quality parameters like DO, salinity, pH, Ammonia, nitrite, microbial load, algal density, *etc.* (Dizge *et al.*, 2011).

Raceway Outlets and Filter Pipes

The drain outlet of raceways is placed in upwards at drain end especially in half way between the end of the partition and the raceway's end wall. The raceway water level can be controlled by an external standpipe positioned inside the harvesting tank. Each raceway should be provided with a different set of perforated filter pipes which are mounted on the outlet to avoid losing fish/shrimp during water drain.

Harvesting Tank

Raceway should be designed to allow drain harvest. Towards the drain end of two raceways single harvesting tank (2 x 1.5 x 2 m) is provided for the purpose of water exchange, bottom waste clearing and for harvesting. Outlet drain pipes of both raceways are connected to harvesting tank and outlet of harvesting tank in turn is connected to the grow out pond. During water exchange for waste removal and harvesting 1000 to 2000μ nylon net screen frame is used inside the harvesting tank (as per the size of shrimps) to avoid escape of shrimps.

Air Lift System

Water circulation in the raceways is depending primarily on airlift systems. In this system, air is introduced 30 cm below the water level in order to attain a maximum uplift of water through a vertical PVC pipe which has a 90° PVC elbow at the upper section. The air will lift the water through the pipe from bottom and sends it through the elbow in desired direction by which unidirectional water flow is ensured. Air supply is ensured by three 5 HP blowers that is to be operated alternatively every 3 hours. Air pressure, airlift pipe diameter, Submergence depth and the type of airlift system being used are among the major factors affecting the pumping rate and water circulated. In the raceway, air lift systems are arranged in 6 sets, 3 on each side of a partition and each set is provided with 3 air lift pipes.

The air lift pumps are fixed to a cross wooden beams with adjustable stainless steel screws which will allow adjusting the height of the airlift pumps in raceway when water level is low.

Rapid Sand Filters

A rapid sand filter with manual backwash and filtration capacity of about $20m^3/h$ is to be provided for each raceways. The sand filter can be used to filter the incoming seawater. The multiport valves in the Rapid Sand Filter is a multi-position valve with six operational, modes sand filter, back wash, rinse, circulation, waste and closed. The backwash and rinse modes serve to maintain the sand filter in optimal working conditions. The circulation bypasses the sand filter. The waste mode can be used to drain the raceway without using the external stand pipe. The closed mode is a safety position to avoid accidental drainage through the multiport valve when the raceway is not in operation (Cohen, 2001).

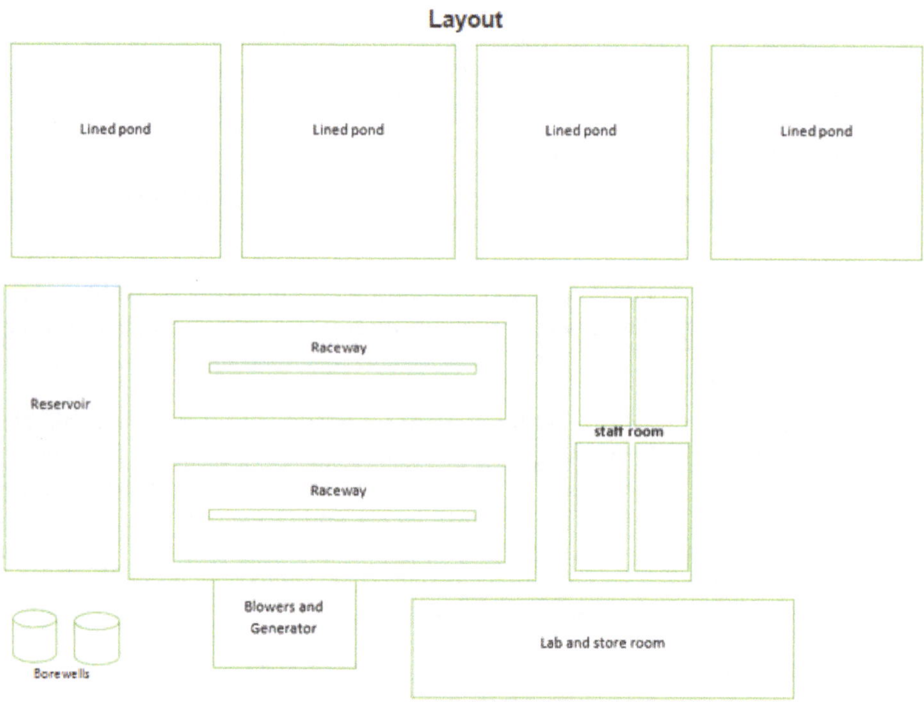

Figure 13.5. Diagrammatic Representation of a Biofloc Farm.

Biological Filter

A conventional gravity flow biological filter is provided for the intake of water. Biological filter tank, having the capacity of 5 x 2 m with central partition and three outlets (50 mm) positioned below the bottom of central partition. In this, one part is provided with different layers of filtration material above the false bottom provided *viz.* big and medium size gravels, smaller size gravels, activated charcoal, coral sand, coarse sand and fine sand. Each layer is separated by a nylon net material. Water will flow through the filtration tank to the other side partition that is into the empty tank and from there to LDPE lined storage pond (Zhu and Chen, 1999).

Figure 13.6. Typical Layout of a Biofloc Farm.

Storage and Mixing Ponds

These ponds are constructed with a dimension of 10 x 5 m each. The dikes and bottom are lined with 250 GSM LDPE sheets to maintain the water quality and to arrest the water seepage. Water from all the bore wells (water of different salinities) is pumped to the mixing pond, where the required salinity is adjusted and from there it is pumped to the biological filter. The bio filtered water is then stored in the storage pond. From there water is pumped to raceways by 5 HP motor through U.V. and rapid sand filters (20,000 LPH).

Blowers, Generator and Paddle Wheel Aerators

Three numbers of twin lobe air blowers (5 HP each) are used in raceways (two are used alternatively and one is kept as stand-by) for the purpose of aeration and operating airlift systems which will enhance the dissolved oxygen level in raceways and also enables continuous water circulation (Boyd and Watten, 1989).

A 15 KVA generator (KIRLOSKAR) is installed to ensure continuous operation of blowers, paddle wheel aerators and sand filters during power failure.

Four numbers of paddle wheel aerators with the capacity of 1 HP are used in each pond. Of them, two are operated at a time for every three hours and afterwards other two aerators are operated. Likewise aerators are operated alternatively every three hours.

Alarm Systems during Power Failure

Stimulation of power failure could immediately activate the alarm system. The alarm sets connected to field lab/control room, which is to be placed at the center of farm site. When operating personals/technicians get the power failure alarm signals, they can switch on the generator immediately without any delay to protect the culture animals.

References

1. Boyd, Claude E., and B. J. Watten. "Aeration systems in Aquacultural." (1989): 425-472.
2. Cohen Y (2001) 'Biofiltration – the treatment of fluids by microorganisms immobilized into the filter bedding material: a review', Bioresource Technology, 77, 257–274.
3. Dizge N, Tansel B and Sizirici B (2011) 'Process intensification with a hybrid system: A tubular packed bed bioreactor with immobilized activated sludge culture coupled with membrane filtration', Chemical Engineering and Processing: Process Intensification, 50, 766–772.
4. Felix, S., 2008. Bio-secured shrimp raceway technology in bio-secured aquaculture: Principles and prototype. Published by AGROBIOS.India. 71-134.
5. Felix, S., 2012. Shrimp Raceway Aquaculture Technology for India, 2012, Narendra Publishing House, New Delhi. Ist Edition. Pg. no. 23.
6. Felix, S., 2014. Aerobic microbial fioculant (AMF) based nursery raceway system and lined ponds for improving Vannamei shrimp production in India-TNFU's initiatives. In sustainable shrimp farming way forward, Aquaculture Chennai -2014
7. Woods, P.J. 1996 - Flood Protection Dykes and Environmental Concerns (Water Resource Branch).
8. Helfrich, L. A., and Garland B. Pardue, B. 2009. Pond construction: some practical considerations. Virginia Cooperative Extension publication No. 420-011, P.1-4
9. Zhu, S. and Chen S., 1999. An Experimental Study on Nitrification Biofilm Performances Using a Series Reactor System. Aquacult. Eng. 20, 245-259.

Chapter 14

Recirculatory Aquaculture Systems

Ramesh Rathod

Scientist, Regional Research Centre, ICAR-CIFA,
Poranki P.O., Vijayawada – 521137, A.P

Recirculating aquaculture systems (RAS) are systems in which aquatic organisms are cultured in water which is serially reconditioned and reused. These systems have minimal exchange of water during a production cycle, hence they are called as 'closed' systems. The main features are limited water exchange, rigorous monitoring and control of water quality parameters, high stocking densities, and high yield per unit volume of water. Physical, chemical and biological treatment of water is crucial in RAS. Hybrid striped bass, channel catfish, and tilapia are the most common freshwater species grown in these intensive culture systems.

Recirculatory System Applications

- ☆ Broodstock maturation
- ☆ Larval rearing systems
- ☆ Nursery systems
- ☆ Nutrition and health research systems
- ☆ Short-term holding systems
- ☆ Ornamental and display tanks
- ☆ High density grow out of food fish

Through water treatment and reuse, recirculating systems utilize less than 10 per cent of the water required by ponds to produce similar yields. To provide high growth, fish are fed high-protein pelletized diets at rates ranging from 15 to

1.5 percent of their body weight per day depending upon their size and species. Aquaponics is the combined culture of fish and plants in recirculating systems.

Components Used in Recirculating Aquaculture Production Systems

Culture Area (Tank)

This area is used to house culture species. Popular tank shapes are either round or rectangular. However, the ideal tank shape depends on the aquatic species being cultured.

Generally, round tanks offer better water circulation compared to the rectangular tanks since later one creates dead spots. Think about the shape of a toilet bowl and the movement of water that happens when you flush. The same water flow is used in round tanks to remove solids from the culture area. This also explains why round tanks often have center standpipes. Rectangular tanks are also used in RAS, particularly for crustacean culture. These tanks offer better use of available space. Water flow in rectangular tanks is typical an inlet (located on one of the short ends) and a standpipe (located on the opposite short end). This design can create dead spots in the corners of rectangular tanks.

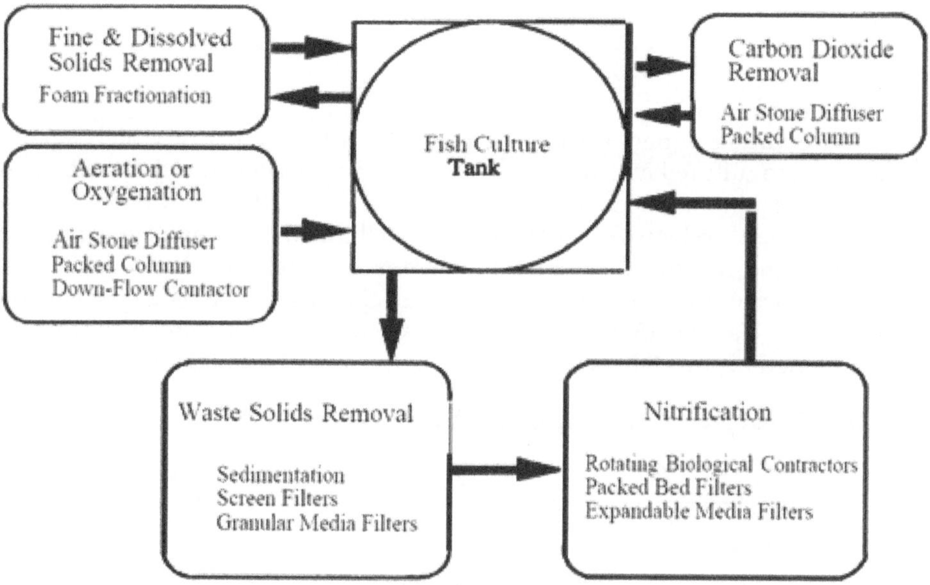

Figure 14.1. Processes Involved in RAS.

Solid Removal

This area is used to remove solids from RAS. Solids are classified as settleable and unsettleable. Settleable solids are larger and weigh more than unsettleable solids. As a result, gravity and mechanical filtration are popular methods of reducing settleable solids. Due to less density, unsettleable solids stay suspended in the water. Removal of unsettleable solids is more difficult than settleable solids. Mechanical filtration is one method of reducing unsettlable solids. Settling Basins are commonly used to remove settleable solids from RAS. Such basins remove solids by slowing down water flow (which can keep settleable solids suspended) and allowing gravity to pull solids to the bottom of the basin. These solids can be removed from the system by flushing the settling basin, which concentrates the solids into sludge, minimizing the amount of water loss.

Mechanical Filtration includes all substances used as screens to prevent certain-sized solids from passing through. Mechanical filtration is useful for removals of both settle able and unsettle able solids but care should be taken to ensure screening materials do not clog too easily. To maintain mechanical filters, the filter material should be removed and cleaned often. Most current technologies for removing suspended solids generally involve some form of mechanical filtration. Two types of mechanical filtration are screen filtration and expandable granular media filtration.

Screen Filtration

Screen filters use some form of fine mesh material (stainless steel or polyester) through which effluent passes while the suspended solids are retained on the screen. Solids are usually removed from the screen by rotating the clogged screen surface past high pressure jets of water. The solids are carried away from the screen in a small stream of waste water. The feature that makes each screen filter different and the challenge in designing these units is the process of collecting the solids on the mesh surface. The screening material has been used in a disk configuration, drum screen configuration, and incline belt configuration. In rotating disk filters, water to be treated enters one end of the filter unit and must pass through sequential vertical disks within the filter. A problem with this design is the small amount of screen surface on which to capture solids. In heavily fed production systems, solids can build up so heavily on one side of the filter that the screens collapse from the water pressure. The most common screen filter is the drum filter. With this configuration, water enters the open end of a drum and passes through a screen attached to the circumference of the drum.

Fine and Dissolved Solids

Many of the fine suspended solids and dissolved organic solids that build up within intensive recirculating systems cannot be removed with traditional mechanisms. A process called foam fractionation (also referred to as air-stripping or protein skimming) is often employed to remove and control the build-up of these solids. Foam fractionation is a general term for a process in which air introduced into the bottom of a closed column of water creates foam at the surface of the column. Foam fractionation removes dissolved organic compounds (DOC) from the water column by physically adsorbing DOC on the rising bubbles. Fine particulate solids are trapped within the foam at the top of the column, which can be collected and removed. The main factors affected by the operational design of the foam fractionators are bubble size and contact time between the air bubbles and the DOC. A counter-current design (bubbles rising against a downward flow of water) improves efficiency by lengthening the contact time between the water and the air bubbles. In this design, water is injected into the foam fraction at or through a venturi. The venturi mixes air with the water and the air/water mixture enters the body of the foam fractionators tangentially.

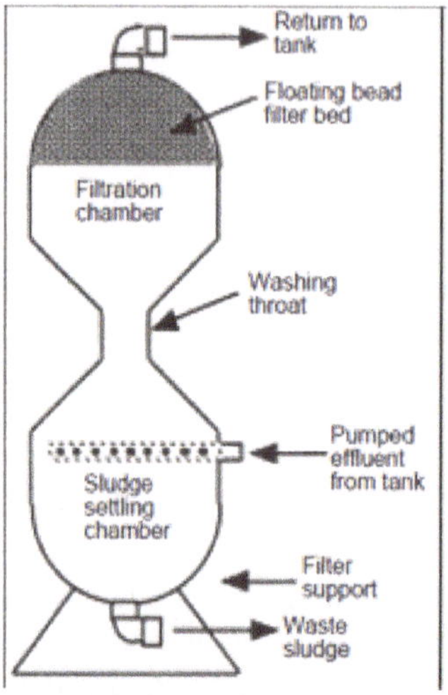

Bubble bead filter

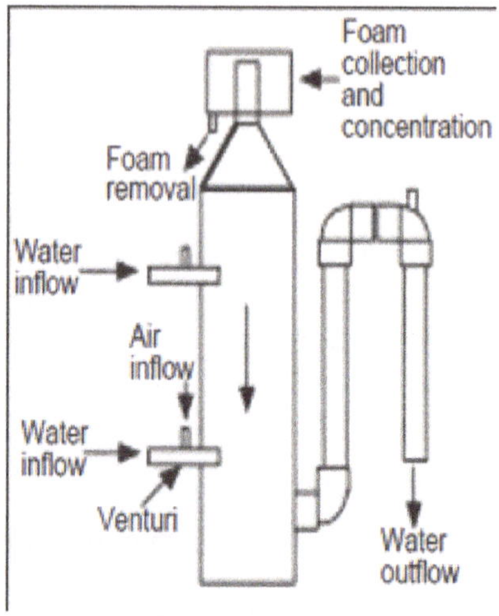

Venturi foam fractionators

Figure 14.2. Type of Screen Filters.

In most installations, the drum rotates only when the filter mesh becomes clogged with solids and a high pressure jet of water (from the outside of the drum) washes the solids off the screen and send into an internal collection trough leading to a waste drain. The advantage of the drum screen filter configuration over the single plate disk filter is the larger surface area of the drum for comparably sized units. The main advantage of using screen filter technology rather than settling basins and swirl separators is their small size and relatively low water loss during backwashing. The main disadvantage of commercial screen filters is cost, especially for smaller units. The smallest commercially available units can process approximately 475 liters per minute (125 gpm) loaded with 25 mg/L of suspended solids. A 100 percent increase in processing capacity increases the cost of a unit by about 50 percent (a unit to process 950 liters/minute). So, larger units are more cost effective. To take advantage of this, the flow streams from several production tanks can be combined into one treatment stream that is cleaned by a larger drum screen filter. However, the advantage of the economy of scale must be weighed against the risk of spreading disease and water quality problems within linked fish production tanks. Vacuum cleaned drum screen filters are now in use. These units have limited capacity (375 to 1,800 L/minute, 100 to 475 gpm) and their performance in commercial facilities has not been well documented. Incline screen or belt screen filters also are beginning to tank. Suspended solids are not be used in the aquaculture industry. These units resemble conveyor belts placed on an incline. Water passes through the screen where suspended solids are retained; solids are lifted out of the water on the incline screen and sprayed off with high pressure water in a cleaning process similar to that of disc and drum screen filters. The units manufactured currently have flow capacities in excess of 7,500 liters per minute (1980 gpm). There is little data on the operational characteristics of these filters.

Biofilter

This area is used to intensify cultures of nitrifying bacteria in RAS. Such bacteria are responsible for removing dissolved ammonia and nitrites. Although nitrifying bacteria live in the water, a biofilter provides additional space for more bacteria to live. As its name suggest, the biofilter is a living component of the system. Both genera of nitifying bacteria (*i.e. Nitrosomonas* and *Nitrobacter*) require adequate food (ammonia and nitrite, respectively), temperature, pH, and dissolved oxygen.

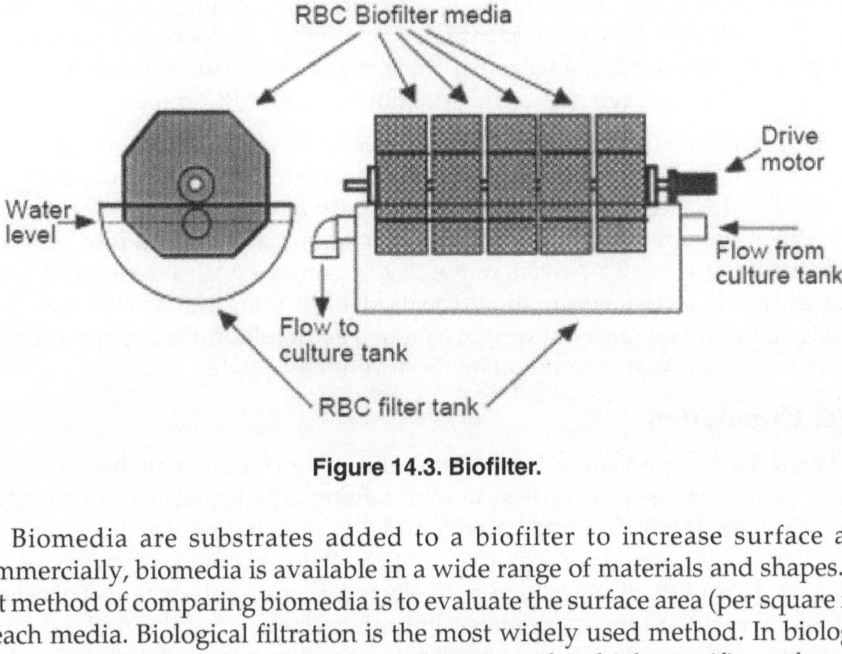

Figure 14.3. Biofilter.

Biomedia are substrates added to a biofilter to increase surface area. Commercially, biomedia is available in a wide range of materials and shapes. The best method of comparing biomedia is to evaluate the surface area (per square foot) of each media. Biological filtration is the most widely used method. In biological filtration (or biofiltration), there is a substrate with a high specific surface area (large surface area per unit volume) on which the nitrifying bacteria can attach and grow. Ammonia and nitrite-nitrogen in the recycled water are oxidized (converted) to nitrite and nitrate by Nitrosomonas and Nitrobacter bacteria, respectively. Commonly used biofilter substrates include gravel, sand, plastic beads, plastic rings, and plastic plates. The most common biofiltration technologies are discussed below.

Rotating Biological Contactor

Rotating biological contactors (RBC) have been used in the treatment of domestic wastewater for decades, and are now widely used as nitrifying filters in aquaculture applications. RBC technology is based on the rotation of a biofilter medium attached to a shaft, partially submerged in water. Approximately 40 percent of the substrate is submerged in the recycle water. Nitrifying bacteria grow on the medium and rotate with the RBC, alternately contacting the nitrogen-rich water and the air. As the RBC rotates, it exchanges carbon dioxide (generated by the bacteria and fish) for oxygen from the air. The tangential velocity of the outer edge of the RBC should be about 35 to 50 feet per minute. For example, an RBC with a diameter of 4 feet would rotate at 3 to 4 revolutions per minute (rpm).

The advantages of RBC technology are simplicity of operation, the ability to remove carbon dioxide and add dissolved oxygen, and a self-cleaning capacity. Major disadvantages are the high capital cost and mechanical instability. Poorly designed or built RBCs can break down mechanically with the weight of the biological growth on the filter medium. RBCs also have been designed to be turned by water (similar to a water wheel) and compressed air. In early aquaculture applications, RBCs had simple discs cut from corrugated fiberglass plate. Now they use media with high specific surface area, such as plastic blocks or a polyethylene tubular medium (resembling hair curlers). These newer plastic media remove more ammonia, nitrite-nitrogen and carbon dioxide in small RBC units.

The plastic media have specific surface areas of up to 200 m^2/m^3 (69 ft^2/ft^3). In aquaculture applications, volumetric nitrification rates of approximately 76 g TAN/m^3 per day can be expected with this type of biological filter (Wheaton *et al.*, 1994). When including these filters in a recirculating system as a nitrifying filter component (assuming 2.5 percent of the feed becomes TAN), a design criterion of 3.6 kg feed/day/m^3 of medium should be used (0.189 pounds/day/ft^3 of medium). The filter medium increases in weight as much as 10 fold during operation, so the support structure must accommodate the additional weight.

Water Circulation

This area is responsible for maintaining water movement throughout the system. Such movement is critical to maintaining aquatic species and nitrifying bacteria in RAS. Pumps are often used for this purpose but are not the only option. Air lifts can also be used for water circulation. Pumps come in a variety of sizes and types. However, all pumps require electricity and replacement. This should be considered at the time of purchase and replacement parts should be on hand. The system will not survive the time it takes for pump parts to be ordered and arrive. Air Lifts are another method of water circulation. Air is used to move water from one component to the next. Air lifts have limited head distances but are quite useful in many RAS. In addition to water circulation, air lifts also provide aeration to RAS waters.

Aeration

Dissolved oxygen is critical to the health of aquatic species and nitrifying bacteria. Oxygen must also be constantly replaced since it is constantly being used. In RAS, aeration is provided with air compressors, air line, and air stones. Air stones play an important role in aeration of RAS. Air from the compressor is forced through an air stone, which is made from a porous material. As a result, air is broken into smaller bubbles and releases oxygen into the water as the air bubble rise through the water. Smaller air bubbles release more oxygen into the water. So the most efficient air stones are those producing finer bubbles. Recirculating systems should maintain adequate dissolved oxygen (DO) concentrations of at least 6 mg/L and keep carbon dioxide (CO_2) concentrations at less than 25 mg/L for best fish growth.

Oxygenation

Pure oxygen is used in recirculating systems when the intensity of production causes the rate of oxygen consumption to exceed the maximum feasible rate of oxygen transfer through aeration. Sources of oxygen gas include compressed oxygen cylinders, liquid oxygen (often referred to as LOX), and onsite oxygen generators. In most applications, the choice is between bulk liquid oxygen and an oxygen generator. The selection of the oxygen source will be a function of the cost of bulk liquid oxygen in your area (usually dependent on your distance from the oxygen production plant) and the reliability of the electrical service needed for generating oxygen on-site. Adding gaseous oxygen directly into the culture tank through diffusers is not the most efficient way to add pure oxygen gas to water. At best, the efficiency of such systems is less than 40 percent. A number of specialized components have been developed for use in aquaculture applications.

U-tube Diffusers

At high operating pressures, more oxygen can be absorbed by water. A U-tube oxygen diffusion system is an energy efficient method of adding pressure to a flow-stream. A typical u-tube consists of a contact loop, usually a pipe within a pipe, buried in the ground to at least 10 meters (33 feet), the height of water required to add one atmosphere of pressure (1 bar,14.7 PSI). The contact loop is placed below tank level to minimize energy requirements, rather than pumping water uphill to gain the extra hydrostatic pressure created by a column of water. Oxygen is mixed with the water at the entrance to the u-tube and travels with the flow to the bottom of the water column. The additional pressure from the water column accelerates the rate of oxygen absorption into the water. The principal advantages of this system are the low energy requirements for oxygenating large flow-streams and the resistance to clogging with particulate solids. The major disadvantage is the construction cost of drilling the shaft and installing the u-tube. Oxygen transfer efficiencies are generally below 70 percent, with effluent oxygen concentrations of up to 250 percent of atmospheric saturation (15 to 20 mg/L).

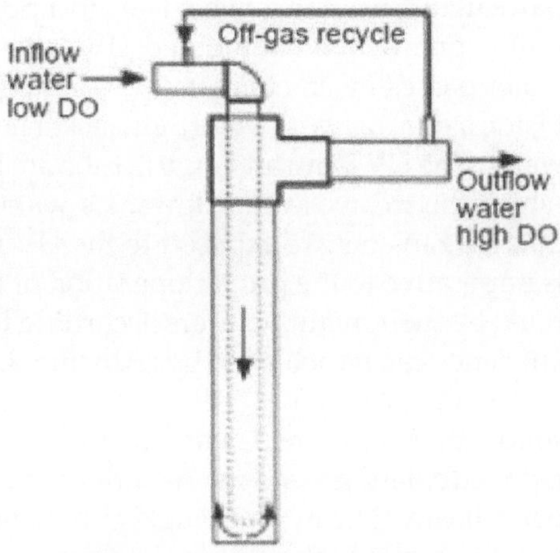

Figure 14.4. U-Tube Diffuser.

Pressurized Packed Columns

Pressurized packed columns are usually operated in a flooded mode (water fills the reactor). Water enters the top of a pressurized chamber that contains a medium with a high specific surface area (much like packed towers). Oxygen gas is usually introduced at the bottom of the column and travels upward, counter to the water flow. Oxygen transfer efficiency can range from 50 to 90 percent with effluent dissolved oxygen concentrations in excess of 100 mg/L. The major disadvantages of this system are high energy requirements (to provide the pressure) and the buildup of biological growth on the packing medium, which makes periodic cleaning necessary.

Ultraviolet Irradiation

Microorganisms (including disease-causing bacteria) are killed when exposed to the proper amount of ultraviolet (UV) radiation. Spotte (1979) reported that the effectiveness of UV sterilization depends upon the size of the organism, the amount of UV radiation, and the level of penetration of the radiation into the water. To be effective, microorganisms must come in close proximity to the UV radiation source (0.5 cm, 0.2 inches or less). Turbidity reduces its effectiveness. For keeping the UV radiation system effective, the water should be pre-filtered with some form of particulate filtration device. The most popular and effective type of UV sterilization unit is one with a submerged UV radiation source. In this type of unit, recycled water passes by an elongated UV lamp (much like a neon light bulb). The lamp is inside a quartz glass, watertight jacket and does not come in direct contact with the water. The UV lamp and quartz tube are held within a small diameter pipe through which the treated water flows. As water passes along and around the UV lamp, microorganisms are exposed to the UV radiation. Keeping the quartz jacket clean is imperative to the proper operation of the unit. UV sterilization units are usually rated by their manufacturers according to their water flow rate capacity. Increased efficiency can be achieved by reducing the flow rate through a given unit. The main disadvantage of UV sterilization is the need for clean water with low suspended solids concentrations. Clear water is not always economically achievable in heavily fed recirculating systems. Additionally, the expensive UV lamp must be replaced periodically. The main advantage of UV sterilization is that it is safe to operate and is not harmful to the cultured species.

Ozonation

Ozone (O_3) gas is a strong oxidizing agent in water. Ozone has been used for years to disinfect drinking water. However, because of the high levels of dissolved and suspended organic materials in recirculating systems, the effect of ozone on bacterial populations is questionable (Brazil *et al.*, 1996). The efficiency of the disinfecting action depends upon the contact time and residual concentration of O_3 in the water with the microorganisms. Ozone must be generated on-site because it is unstable and breaks down in 10 to 20 minutes. Ozone is usually generated with either a UV light or a corona electric discharge source. There are many commercial ozone generation units available. Ozone is usually diffused into the water of a recirculating system in an external contact basin or loop. Water must be retained in this side-stream long enough to ensure that microorganisms are killed and the ozone molecules are destroyed. Residual ozone entering the culture tank can be very toxic to crustaceans and fish. Ozone in the air is also toxic to humans in low concentrations. Great care should be taken in venting excess ozone from the generation, delivery, and contact system to the outside of the building. Ozonation systems should be designed and installed by experienced personnel.

Heating and Cooling

This area is responsible for meeting the temperature requirements of aquatic species. Heating and cooling of RAS water is often required to raise exotic animals. But both processes require large amounts of energy to do and marketing research should be performed to justify these expenses. In Kentucky, heaters are more commonly used than chillers.

Advantages of RAS

- ✰ Rigorous control is possible
- ✰ Unaffected by weather
- ✰ Predators and competitors are easily excluded
- ✰ Efficient feeding
- ✰ Requires minimal water
- ✰ Minimal environmental impacts

Disadvantages of RAS

- ✰ High capital costs
- ✰ High running costs (pumping, filtration, *etc.*)
- ✰ Requires very careful management
- ✰ A pathogen outbreak will be severe

References

1. Acierno, R., 2006. Manual on Effluent Treatment in Aquaculture: Science and Practice. Outcome of the EU supported Aquatreat.org project, 2007, P.163.
2. Alfred Jokumsen and Lars M. Svendsen. 2010. Farming of Freshwater Rainbow Trout in Denmark. DTU Aqua, National Institute of Aquatic Resources. DTU Report no. 219-2010. ISBN 978-87-7481-114-5
3. Aquaculture, Volume 1 and 2, Edited by Gilbert Barnabé, Ellis Horwood Limited, Chichester, West Sussex, PO19 !EB, England, 1990, ISBN 0-13-044108-2
4. FAO, 2014. The State of World Fisheries and Aquaculture 2014, FAO Fisheries and Aquaculture Department, Viale delle Terme de Caracalla, 00153 Rome, Italy, 2007, ISBN 978- 92-5-108275-1
5. Fredrick W. Wheaton, 1993. Aquacultural Engineering by, Krieger Publishing Company, Malabar, Florida, 32950 USA, ISBN 0-89464-786-5

6. James W. and Avault Jr., 1996. Fundamentals of Aquaculture, A Step-by-Step Guide to Commercial Aquaculture by AVA Publishing Company Inc., Baton Rouge, Louisianna 70884-4060 USA, ISBN 0-9649549-0-7
7. Lydia Brown, 1993. Aquaculture for Veterinarians: Fish Husbandry and Medicine, Pergamon Press Ltd., Oxford, UK, ISBN 008-040835
8. Remmerswaal, R.A.M., 1997. Recirculating Aquaculture Systems by, INFOFISH Technical Handbook 8, ISBN 983-9816-10-1
9. Thomas D. B., David W. S. and Michael T., 1984. Madigan,Biology of Microorganisms by Prentice-Hall International, USA, ISBN 0-13-078338-2
10. Timmons, M.B and Ebeling, J.M., 2002. Recirculation Aquaculture by NRAC Publication No. 01-007, Cayuga Aqua Ventures, USA, ISBN 978-0-9712646-2-5

Chapter 15

Aquaponics: A New Emerging Aquaculture Technology Integrated with Hydroponics

A. Devivaraprasad Reddy[1], *Ch. Bala Krishna*[2], *T. Vijaya Nirmala*[1], *E. Karuna Sree*[1] *and D. Chinnam Naidu*[2]

[1]ICAR-Krishi Vigyan Kendra, Dr. YSR Horticultural University, Venkataramannagudem, West Godavari District, Andhra Pradesh
[2]ICAR-Krishi Vigyan Kendra, Acharya NG Ranga Agricultural University, Amadalavalsa, Srikakulam, Andhra Pradesh

Population in the world is increasing day by day which triggers more food production through sustainable intensification strategies and the pressure on the resources like land, water and nutrients though they remain same. To address these issues there is a vital alternative suitable, ecofriendly, reliable and sustainable methods to produce food. Aquaponics, a technology that bio-integrates the two systems *i.e.* aquaculture with hydroponics (Vegetable, flower, herbs production *etc.*), which is gaining more attention towards bio-integrated food production system with effective utilization of space and resources. Aquaculture is the cultivation of aquatic organisms (fish, crustaceans, molluscans *etc.*) or plants (seaweeds, freshwater macrophytes *etc.*) which are living in water for production. Farming of food materials in an inert medium by addition of nutrients without any soil under controlled conditions is otherwise called Hydroponics. Whereas in aquaponics, growing of fish in a confined area where there is no effect of substances and the accumulation of nutrient-rich waste water is produced and act as a source of natural fertilizer for the growing plants. These plants purify the water by way of consumption of the nutrients being produced in the aquaponics. This is also the answer for using marginal lands in the urban areas, since; aquaponics utilize the waste water and reusing it. Hence this can also be used as backyard production units or terrace gardening. Fish farmers who are facing in difficult for disposing the waste water and also to increase the productivity by growing the plants by way of hydroponic unit is the best alternative for adopting aquaponics.

Principles of Aquaponics

Aquaponics is having many principles and basically works as an ecofriendly way of food production systems. The end product of one system *i.e.* processed water produced from one system will serve as raw material product for the second system. Where the resources can be effectively utilized and reduce the waste disposal problems. Diversification of high value crops like fish and plants which results in higher productivity and increases revenue generation. The water utilized for the fish production is reused for growing of plants and vice versa by following either coupled or decoupled either single or multiloop systems. The enriched water from the fish tanks is used to fertigate the hydroponic beds.

The microbes present in the roots of plants utilizes the available nutrients from the waste water and purifies the water. The hydroponic beds act as substance for attaching the microbes and reduces the toxic gases available in the water. Thus, the recycling of water takes place and reduces the toxic loads in the water. Ammonia reducing microbes which are present in the hydroponic bed and plant roots play a vital role in nutrient cycling.

History of Aquaponics

Around 1000 A.D., Aztec Indians has raised their plants on the surface of a lake using raft as platform though the word Aquaponics was coined during the 1970s. In recent times, aquaponics has become popular in aquaculture since the fish farmers are looking for the alternative methods of farming practices to reduce the available possessions *i.e.* land, water, nutrients and other resources and to increase the income. Aquaculture is rearing of aquatic organisms (fish and shrimp) in earthen ponds or pens or cages. Extensive and semi-intensive culture practices can be done in this man-made pond. From the last few decade onwards people started using the recirculating aquaculture systems (RAS). In this RAS, high stocking density, reusing of water reduces and limited land utilization could be possible. However, the huge amount of waste water is quickly accumulating in this system and it is highly rich in nutrients too. University of Virgin Islands, started utilizing this waste water for growing of plants as a part of research program during 1970s. In the mid-1980s, the first commercial aquaponics was established at Bio-shelters in Amherst, MA. Home based systems as backyard aquaponics started using this technique as smaller systems and also refined the growing media bed for growing as backyard aquaponics for smaller systems.

Aquaponics - How it Works

Plants synthesize their food by utilizing the water, light, CO_2, trace elements *etc.* In lack of any of these elements, plants will either withstand or die. Fish waste water is usually rich in ammonia and when it builds up in the tank and results in death of fish. Ammonium (NH_4^+), nitrite (NO_2^-) and nitrate (NO_3^-) are the foremost common ionic (reactive) types of dissolved inorganic nitrogen in aquatic ecosystems. These ions are gift naturally as a result of atmospherically deposition, surface and groundwater runoff, dissolution of nitrogen-rich geologic deposits, N_2 fixation by some prokaryotes (cyanobacteria with heterocyst's), and biological degradation of organic matter.

Nitrification and denitrification process are performed by the ammonia oxidizing microbes including ammonia oxidizing archaea (AOA) and ammonia oxidizing bacteria (AOB) *viz.*, Nitrosomonas, Nitrobacter consume ammonia and give off nitrites, which Nitro-bacterium eat and excrete nitrates. This process consumes heterotrophic bacteria as well as autotrophic microbes. Heterotrophs consume oxygen, carbon, inorganic nutrients and generates nitrate. The pH lowers as the nitrifying bacteria multiply. If the water circulates the entire tank, then water carries the equal parts of nutrients and toxic gases if any. If the water colour is turbid then reduction of feeding ration for some days will solve the problem. The plant consumes nitrates that produced from the waste water. Basically, fish produces the ammonia and the bacteria grown on the beds can consume that ammonia and produces nitrates.

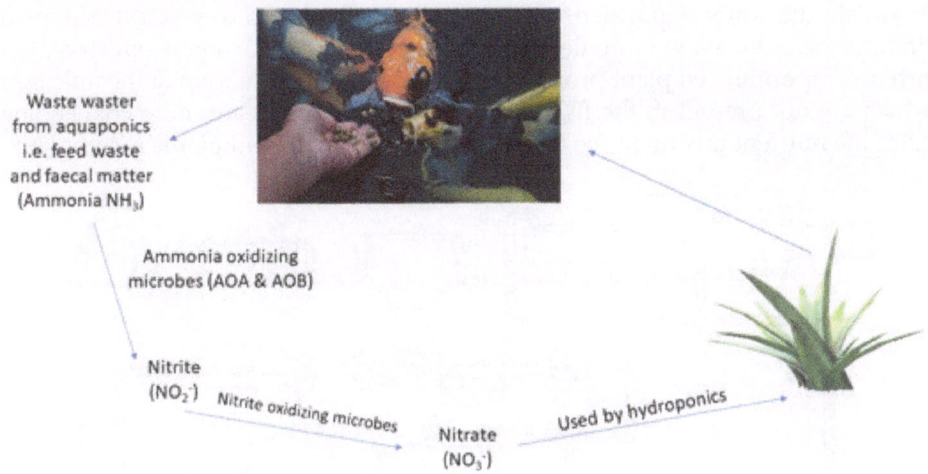

Figure 15.1. Working Principles of Aquaponic System.

The plant utilizes nitrates to produce food and the water circulates the entire system, which is cleaner and oxygenated. Fish and plant produce can be eaten by humans and left-over plants used to compost by using earthworms to decompose. If the water ciruculation is stopped, it will lead to accumulation of nitrates, low dissolved oxygen consequently formation of stratification. Thus the water circulation reduces many problems. Usage of water depends on several factors which are inherent in the system. Traditional farming consumes 70-95 per cent water while 20-30 per cent of water only consumed in this system. Two important factors that consume more water in the aquaponics system *i.e.* evaporation and vegetable plants. Chlorinated water should not be used in the systems since it causes adverse effects on cultured animals. Chlorinated water should be used after for more than 24 hours allowing it for settlement and evaporation of chlorine residues. Bioreactors can also be utilized for sludge removal and recycling of organic wastes. Solar power can also be used for more energy efficient operation of aquaponics equipments. The unique solution in the aquaponics, is to recycle the water which reduces the environmental problems.

Aquaponic Designs

Aquaponics is having three different designs *viz.*, coupled (one loop) systems, decoupled (two loop) systems, and decoupled (multi-loop) system. In the coupled (single loop) system, the processed water moves from aquaculture to hydroponics unit and vice versa without any mineralization or filtration steps. Decoupled (two loop) system, the processed water flows from aquaculture to hydroponic unit only. Whereas in the decoupled (multi-loop) system the process water passages from aquaculture to hydroponic unit following through various processing steps like alkalization, mineralization *etc.* Decoupled aquaponic designs works on the principle that the water enriched with nutrients produced in the fish culture tank, are used to grow the plants and in turn the water sent back to the fish culture tank. In which, the water is purified without much recirculation of water. Decoupled designs therefore allow more flexibility in customizing the water chemistry, after fish use, for optimized plant production because supplementation of the nutrients which are not present in the fish feed. Decoupled designs are more challenging since the nutrient mixing in the water for growing of fish and plants is complex.

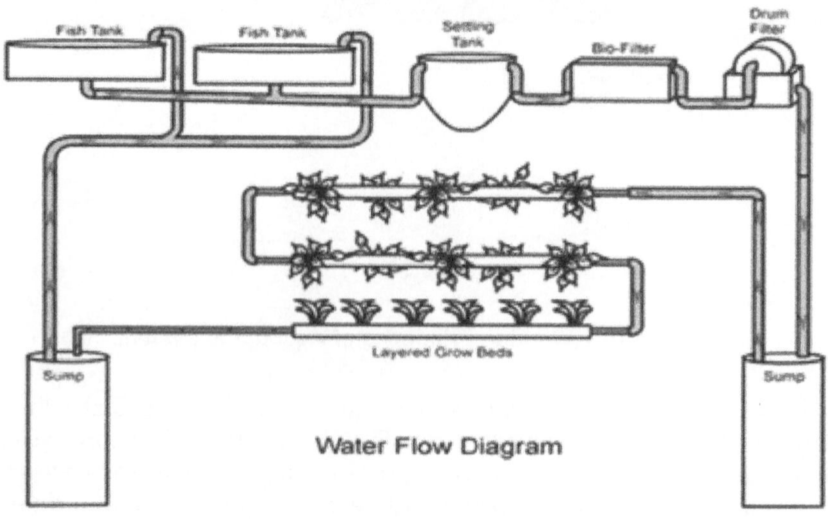

Figure 15.2. Design of Aquaponics.

Aquaponics Systems

Aquaponic systems are classified into mini, hobby, backyard, semi commercial and large scale or commercial system. Parts of Aquaponic System Aquaponics system comprises of two main parts *viz.*, aquaculture and hydroponics. Intensification of stocking density, left over feed, faecal matter accumulates in the closed aquaculture systems. This ammonia rich water becomes more toxic to the fishes living in the water. The plants utilize some of the nutrients which are vital for growth. There are numerous systems or sub systems are there in aquaponics which are useful for exclusion of solid wastes.

Typical components of Aquaponics include the following (1) Fish Tank, (2) Media - Place to grow plants, (3) Water pump, (4) Air pump, (5) Irrigation tubing, (6) Sump, (7) Settling basin, (8) Biofilter, (9) Water Temperature (optional), (10) Grow light (optional) Fish Tank/Rearing tank.

Fish Tank/Rearing Tank

Fish rearing tank is one of the most significant mechanisms of any aquaponics system and the size and shape of the tanks varies with requirement. Polythene lining sheet tanks work better with less problems than the earthen ponds. The slope of the floor would be either one side of the corner or center of the tank for the easy way to collect, clean the waste from the tank. The main important things to be considered for designing of system are selection of suitable fish and plants for growing and the culture technologies should not intrinsically uptake or harmfully utilize the water or nutrient resources added in the system. For example, earthen ponds are not preferred for growing of fish, since these ponds which influence for water unavailability and also affect the nutrient resources for growth of fish and plants. Likewise, hydroponic systems for growing of plants should not use of any media that influences on the uptake of excessive nutrients or water.

Plant Production Systems

Plant production systems are many types within revised agriculture float technology (RAFT) *viz.*, Deep water culture, Media based and Nutrient film technique (NFT). Further it is classified into two main clusters: i) NFT is used as a media for growth of plants for their anchorage and also act as a substrate for the growth of microbes for nutrient cycling. ii) Modified design in the plant production system *viz.*, Vertical farming, Robotics, and Complete artificial light. These last systems can be further distinguished on the basis of substrate used: organic, inorganic and synthetic. The farming of many varieties of plants, herbs, fruiting crops and leafy vegetables can be grown. In many cases, the nutrient content of the process water in the aquaponics are sufficient for good plant growth.

Operation

Always use the non-chlorinated water and fill the tank or sump using pump. Lift the water into the system using the power assisted pump. This water flows from one place to another place once it completely fills to the marked point and thus recirculation takes place. Cycle depends on the system design and size. Usually it takes 20-30 minutes long and afterwards using automatic power assisted switch will turn on and off at a regular interval. Always check the water retention in the beds. It should not be empty. This provides moisture to the beds where the plant roots absorb the nutrients and the microbes that also takes advantage of this and start working on it. By any chance, the pump failure that does not harm to the plants for about a day, but it harms the fish. Since high density of fish are being stocked and the accumulation of the toxic gases creates suffocation to the fish and slowly starts dying. Thus, the system should be checked visually whenever possible. Water level can be adjusted based on the retention time in the bed and delivery line.

Plants Adapted to Aquaponics

Crops growing under controlled conditions using the precise quantities application to the hydroponic plants. It depends on the stocking density of the fish and species. Nevertheless, in aquaponics, the waste water generated from the aquaculture is being used for growing the plants. This enriched waste water is managed by using the feed and faecal waste. This may contain sufficient levels of ammonia, nitrate, nitrite, phosphorus, potassium, and other secondary and micronutrients to produce hydroponic plants. Lettuce, herbs, and specialty greens (spinach, chives, basil, and watercress) have low to medium nutritional requirements and are well suitable to the aquaponic systems. Plants yielding fruit (tomatoes, bell peppers, and cucumbers) have a higher nutritional demand and perform better in a heavily stocked, well established aquaponic system. Greenhouse varieties of tomatoes are better adapted to low light, high humidity conditions in greenhouses than field varieties.

Fish Species

Fish species like tilapia, pangas, trout, perch *etc.* which are adapted to recirculation aquaculture systems are more suitable fish for aquaponic system. In most of the commercial based aquaponic systems, tilapia is preferred since the fish is highly tolerant to fluctuating water conditions such as pH, temperature, dissolved oxygen, dissolved solids *etc.* Common carp is also suitable for growing in this system.

Water Quality Characteristics

Water is the elixir of life and is becoming increasingly scarce nowadays. Hence, the recycle and reuse of water is the need of the hour. Water consumption in the aquaculture industry and discharge of wastewater from such an industry is an ecological concern world-wide. Disposal of waste water from aquaculture industries presents a major problem because of their objectionable odour, high protein and residual ammonia content. The importance of maintaining quality and complying with regulations regarding liquid effluents that are becoming more stringent which cannot be over emphasized. Recirculatory aquaculture systems require good water quality parameters for their growth and life. They are pH, dissolved oxygen, alkalinity, hardness, salinity, carbon dioxide, chlorine, and other parameters. The water quality parameters depend on the stoking density, size of fish stocked, feeding ration *etc.* In addition to these, environmental factors also influence the water quality. Water quality monitoring is more crucial for growth and condition of the fish. If healthy environment is present, better growth can be expected with no or minimal disease incidence.

Biofiltration

Waste water from the aquaculture contains ammonia, nutrient rich and high in dissolved solids. Some of the aquaponics systems will use the intermediate filtration units for reducing the suspended solids from the waste water. Sometimes these suspended solids create problem in water circulation and also disturb the plant absorption. For better circulation of the water, intermediated filter cleaning and backwash will maintain the system for long system without much problems. Gravel in the hydroponic unit acts a substrate for attaching the microbes and functions as "fluidized bed bioreactor,". It eliminates the dissolved solids and making the habit for better growing of nitrifying bacteria which is in nutrient conversions.

Component Ratio

Ration of hydroponic media with the volume of aquaculture tank is nothing but component ratio and usually the ration between tank:bed is 1:1 or 1:2 is generally implemented in early days and 1:2 is commonly used. But, recently it has been raised to 1:4 or even 1:6. However, this ratio varies with various factors like stocking density of the fish, fish species, type of crop grown, feeding rate, *etc.*

Guidelines for Successful Aquaponic System Maintenance

- ☆ Use of correct feeding rate ratio for design calculations
- ☆ Keeping feed input relatively constant
- ☆ Supplementation with calcium, potassium and iron
- ☆ Ensuring good aeration
- ☆ Removal of solids
- ☆ vigilance with aggregates
- ☆ Care on use of oversize pipes
- ☆ Use of biological pest control
- ☆ Ensuring adequate biofiltration
- ☆ Controlling pH

Advantages

- ☆ 2 crops from 1 input
- ☆ No effluent (negligible environmental impact)
- ☆ Productivity 6 times higher than soil agriculture and very suitable for islands
- ☆ More efficient and space saving method of growing fish, vegetables and herbs.
- ☆ By incorporating aquaponics, hydroponic growers can eliminate the cost and labor involved in mixing a fertilizer solution and commercial

aquaculturists may be able to drastically reduce the amount of filtration needed in recirculating fish culture

- ☆ Fish culture: 25-30 per cent feed conversion
- ☆ 70-75 per cent excreted as solid or dissolved fractions – accumulates in system
- ☆ Fish water rich in nutrients – for plant growth
- ☆ Plants – biofilters for water regeneration
- ☆ Converts harmful to harmless nitrates and phosphates
- ☆ Nutrient removal by plants to improve effluent quality
- ☆ Utilizes the nutrient rich water from aquaculture that otherwise would have been a waste product.
- ☆ Provides a truly organic, natural form of nutrients for the plants.
- ☆ By eliminating the soil in vegetable production, soil borne disease will be eliminated
- ☆ In aquaponics, plant spacing can be very intensive, allowing to grow more plants in a given space.
- ☆ With high stocking densities in the fish tank, plants will quickly grow and develop in aquaponic system.
- ☆ No use of pesticides or herbicides, making the end product healthier and safer.
- ☆ If climate permits or growing in a greenhouse, crops can be grown year-round in aquaponic system.

Disadvantages

- ☆ Complex equipment necessitates high capital expense and constant electricity
- ☆ Require trained staff for operation and maintenance
- ☆ Initial expenses required for housing, tank, plumbing, pumps, and grow beds.
- ☆ Aquaponics systems can have electrical failure or pipe blockage can lead to a complete loss of fish stock.

The Future of Aquaponics

Latest technologies in the agriculture and allied sectors has improved the productivity during the last century. Nevertheless, they change the ecosystem and produce the food to meet the demand of ever-growing population. Aquaponics has to address the regulatory issues for integrating the technologies. In addition to this, consumer certificate schemes for aquaponics are also a difficult area and in some parts of the world it can be certified as organic produce. Small scale aquaponics units are economically viable where their place of demand for that specific crop and grown in that system. Moreover, in closed recirculatory systems, the risk of pathogen can be lowered and also manage the potential risks. So that food safety can be achieved. These type of aquaponics systems can be used as terrace gardening for production and also as hobby.

References

1. Ako, H., and Baker, A. 2009. Small-scale lettuce production with hydroponics or aquaponics. College of Tropical Agriculture and Human Resources (CTAHR), University of Hawai at Manoa.wwwctahr.hawaii.edu/oc/freepubs/pdf/SA-2.pdf
2. Awad, A. M., 2017. Evaluation of Ammonia (NH3) removal through water recycling in aquaculture by aquaponic system and its effect on the growth rate of used fish and plants. Medrc. MEDRC Series of R and D Reports. MEDRC Project: 16-DC-005
3. By Steve Dive, NCAT Agriculture Specialist, 2006 NCAT, Aquaponics—Integration of Hydroponics with Aquaculture: In National Sustainable Agriculture Information Service, 1-800-346-9140 • www.attra.ncat.org.
4. Connolly, K., and Trebic, T., 2010. Optimization of a Backyard Aquaponic Food Production System. Bioresource Engineering, 1–74.
5. Diver, S., 2006. Aquaponics — Integration of. Water, 1–28. https://doi.org/10.1146/ annurev.genet.37.110801.143717
6. Eck, M., Korner, O., and Jijakli, M. H., 2019. Nutrient Cycling in Aquaponics Systems. In Aquaponics Food Production Systems. https://doi.org/10.1007/978-3-030-15943-6_9
7. FAO Fisheries and Aquaculture Technical Paper No. 589. Rome, FAO. 262 pp
8. Goddek, S., 2019. Aquaponics Food Production Systems. Aquaponics Food Production Systems, June. https://doi.org/10.1007/978-3-030-15943-6
9. Hollyer, J., Tamaru, C., Riggs, A., Klinger-bowen, R., Howerton, R., Okimoto, D., Castro, L., Ron, T. B., Fox, B. K. K., Troegner, V., and Martinez, G. 2009. On-Farm Food Safety/ : Aquaponics. July.
10. https://www.theaquaponicsource.com/what-is-aquaponics
11. Joshi, S. A. K., and Ravindranath, K., 2017. Tilapia - the most amenable fishes introduced and farmed worldwide. Matsya Bharat, Newsletter of the National Fisheries Development Board, 8(5), 26–38. http://nfdb.gov.in/PDF/NEWSLETTERS/15_Matsya Bharat January-March 2017.pdf
12. Lennard, W., 2004. Aquaponics Research at Melbourne Australia. Aquaponics J, 35, 18–24.
13. Maucieri, C., Nicoletto, C., Junge, R., Schmautz, Z., Sambo, P., and Borin, M., 2018. Hydroponic systems and water management in aquaponics: A review. Italian Journal of Agronomy, 13(1), 1–11. https://doi.org/10.4081/ija.2017.1012
14. Rakocy, J.E., Shultz, R.C., Bailey, D.S. and Thoman, E.S. 2004. Aquaponic production of tilapia and basil: comparing a batch and staggered cropping

system. *Acta Horticulturae* (ISHS) 648:63-69 (http://www.actahort.org/books/648/648_8.htm)

15. Resh, H.M. 1995. Hydroponic food production. Woodbridge Press Publishing Company,Santa Barbara, California.
16. Rinehart, L., 2019. Aquaponics – Multitrophic Systems. ATTRA Sustainable Agriculture, April, 23.
17. Sarman, V., and Mahavadiya, D., 2019. Aquaponics/ : Enhance Farmer Income. March, 11–14.
18. Somerville, C., Cohen, M., Pantanella, E., Stankus, A.and Lovatelli, A. 2014. Small-scale aquaponic food production. Integrated fish and plant farming.
19. www.backyardaquaponics.com/
20. www.growingpower.org/education/what-we-grow/aquaponic
21. Yep, B., and Zheng, Y., 2019. Aquaponic trends and challenges – A review. Journal of Cleaner Production, 228, 1586–1599. https://doi.org/10.1016/j.jclepro.2019.04.290

COLOUR PLATES

Figure 1.1. *Penaeus monodon (pg.3)*

Figure 1.2. *Litopenaeus vannamei (pg.4)*

Figure 1.3. *Fenneropenaeus merguiensis (pg.5)*

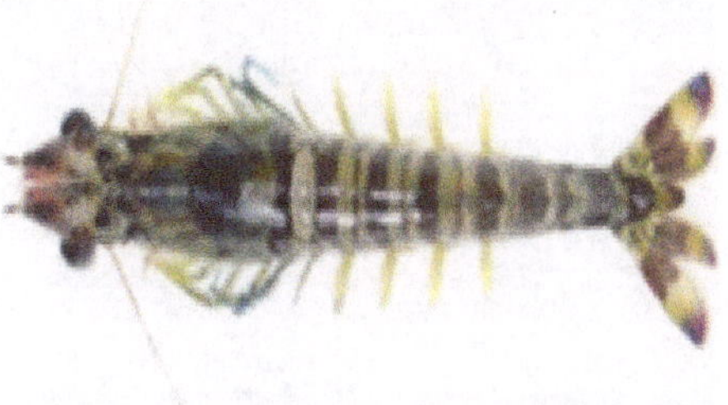

Figure 1.4. *Marsupenaeus japonicas (pg.5)*

Figure 1.5. Fenneropenaeus indicus *(pg.5)*

Figure 1.6. Scylla serrata *(pg.6)*

Figure 1.7. *Lates calcarifer (pg.7)*

Figure 1.8. *Mugil cephalus (pg.8)*

Figure 1.9. *Chanos chanos (pg.9)*

Figure 1.10. *Etroplus suratensis (pg.10)*

Figure 1.11. *Rachycentron canadum (pg.11)*

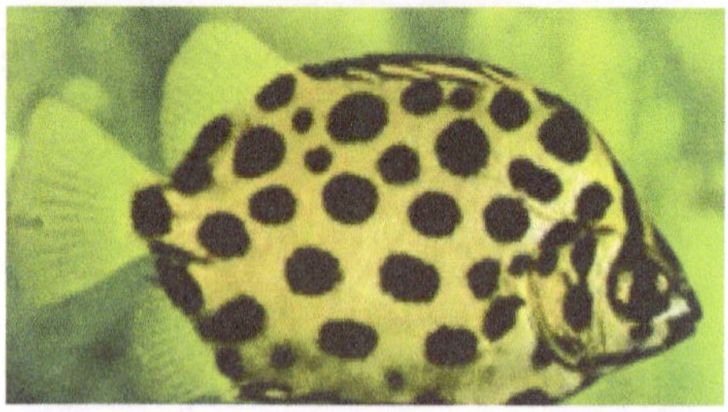

Figure 1.12. *Scatophagus argus (pg.12)*

Figure 1.13. *Monodactylus argenteus (pg.13)*

Figure 2.1. Biofloc Technology (BFT). *(pg.19)*

Figure 3.1. Egg Scatters Laying Non-adhesive Eggs. *(pg.33)*

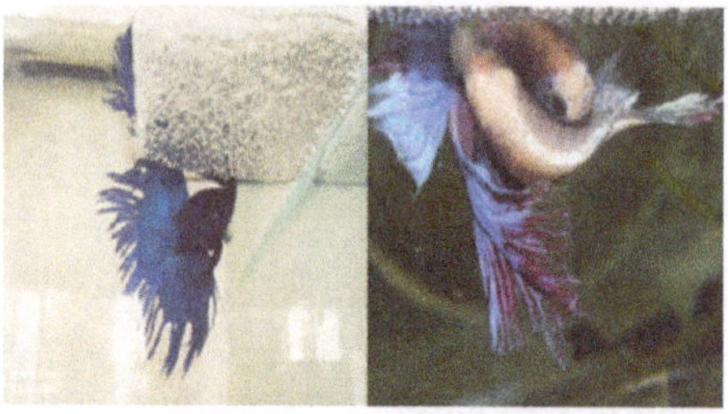

Figure 3.3. Bubble Nest and Mating Betta. *(pg.35)*

Figure 3.2. Betta Breeding Trap. *(pg.35)*

Figure 3.4. Guppy Premating Behaviour and Young Ones. *(pg.39)*

Figure 3.4. Guppy Premating Behaviour and Young Ones. *(pg.39)*

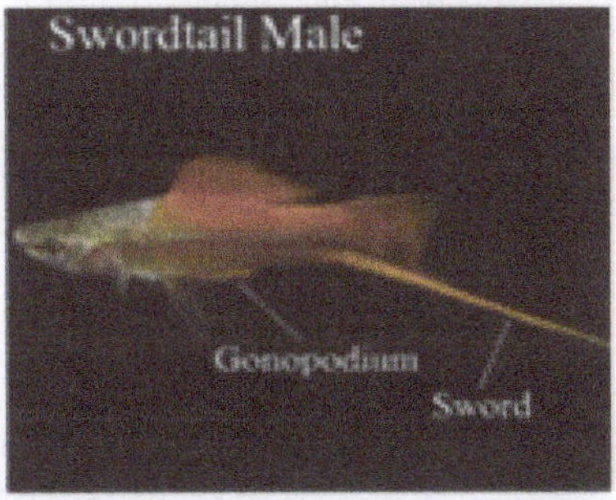

Molly young one

Guppy

Molly

Sword tail

Platy

Figure 3.5. Platy, Swordtail and Molly Fishes. *(pg.40)*

Figure 3.8. Treating with $KMnO_4$. *(pg.45)*

Figure 3.10. Setting of Aquarium. *(pg.49)*

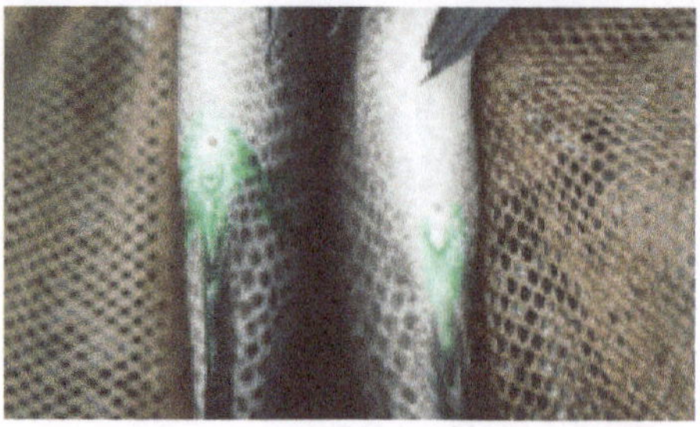

Figure 4.1. Tilapia Male and Female. *(pg. 53)*

Figure 4.3. Broods in Hapa. *(pg. 54)*

Figure 4.4: Egg Collection. *(pg. 54)*

Figure 4.5: Juvenlines of Red Tilapia. *(pg. 55)*

Scylla serrata

Scylla tranquebarica

Scylla paramamosain

Scylla olivacea

Figure 5.1. Mud Crab Species (Courtesy: Keenan *et al.*, 1999). *(pg. 61)*

Figure 4.5: Juvenlines of Red Tilapia. *(pg. 74)*

Figure 6.2. Nursery Rearing in Ponds. *(pg. 79)*

Figure 6.3. Nursery Rearing in Cages/Hapas. *(pg. 80)*

Figure 6.4. Pond Based Culture System. (pg. 83)

Figure 7.2. Reservoir/Settlement Tank. (pg. 91)

Figure 7.3. Catridge Filder and UV. (pg. 91)

Figure 7.4. *Penaeus vannamei*. (pg. 92)

Figure 7.5. *Penaeus monodon*. (pg. 92)

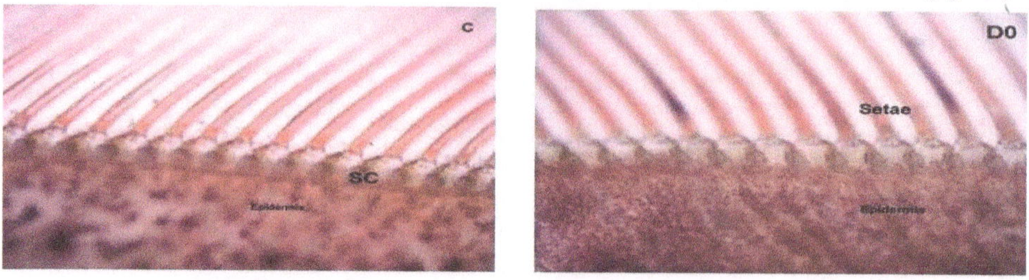

Figure 7.8. Molt Stages in *Penaeus monodon*. (pg. 96)

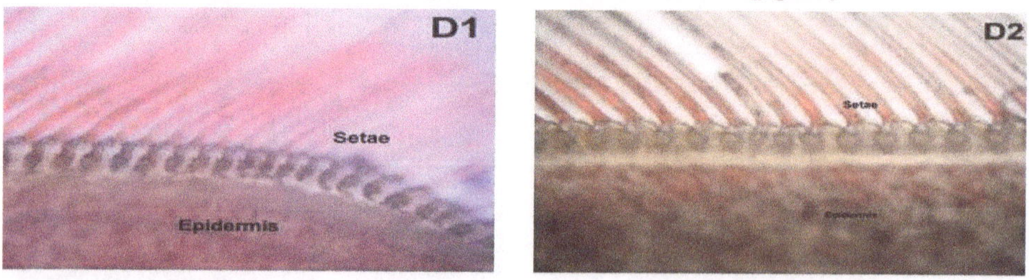

Figure 7.8. Molt Stages in *Penaeus monodon*. (pg. 96)

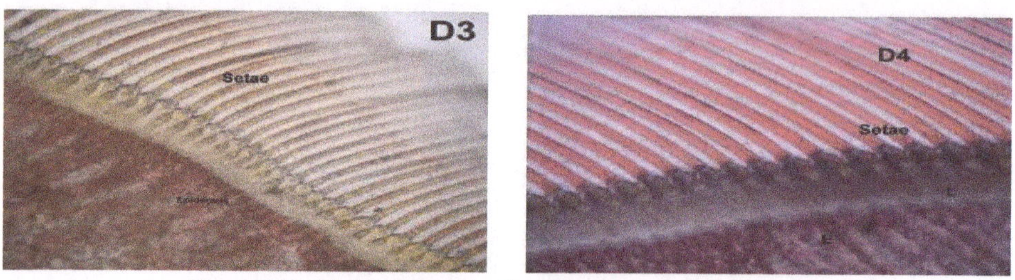

Figure 7.8. Molt Stages in *Penaeus monodon*. (pg. 96)

Top soil removal

Pond bottom after top soil removal

Figure 7.9. Pond Bottom Top Soil Removal. (pg. 103)

Lime applied on pond bottom   Lime application in pond water

Figure 7.10. Lime Application in Pond. (pg. 104)

Figure 7.11. Shrimp Pond Water. (pg. 106)

Figure 8.1. Collection of Polychaete Worms from the Lagoon. (pg. 113)

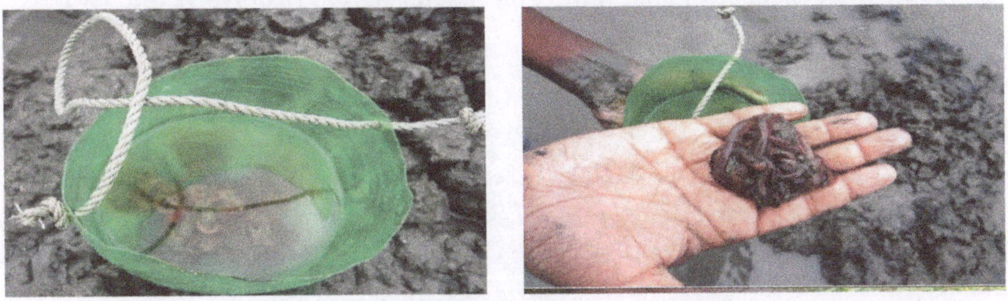

Figure 8.1. Collection of Polychaete Worms from the Lagoon. (pg. 113)

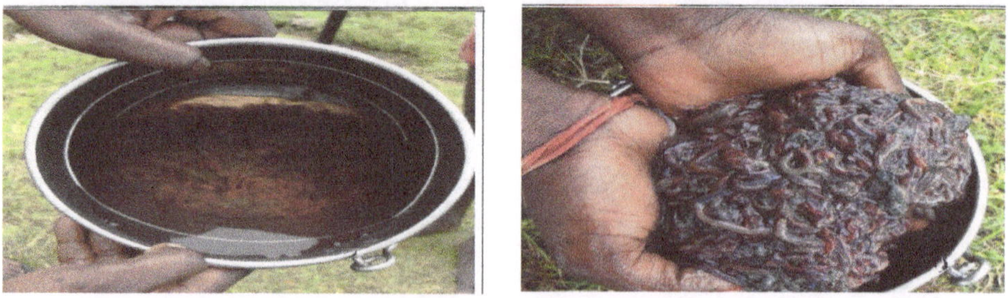

Figure 8.1. Collection of Polychaete Worms from the Lagoon. (pg. 113)

Figure 8.2. Development of Tanks with the Substrate for Grow-Out Practices of Mud Worm Polychaete, *Marphysa gravelyi*. (pg. 114)

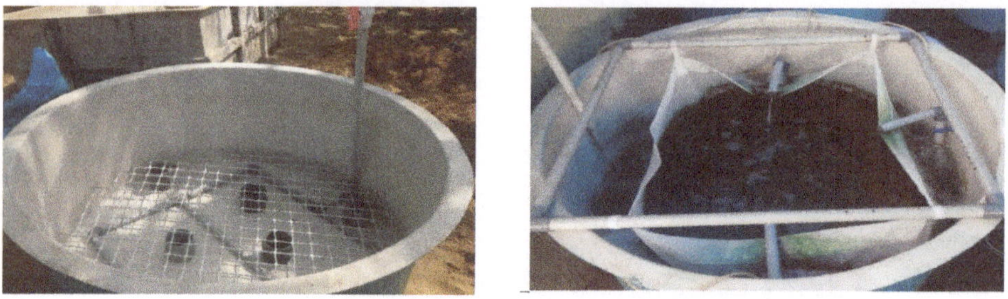

Figure 8.2. Development of Tanks with the Substrate for Grow-Out Practices of Mud Worm Polychaete, *Marphysa gravelyi*. (pg. 114)

Figure 8.2. Development of Tanks with the Substrate for Grow-Out Practices of Mud Worm Polychaete, *Marphysa gravelyi*. (pg. 114)

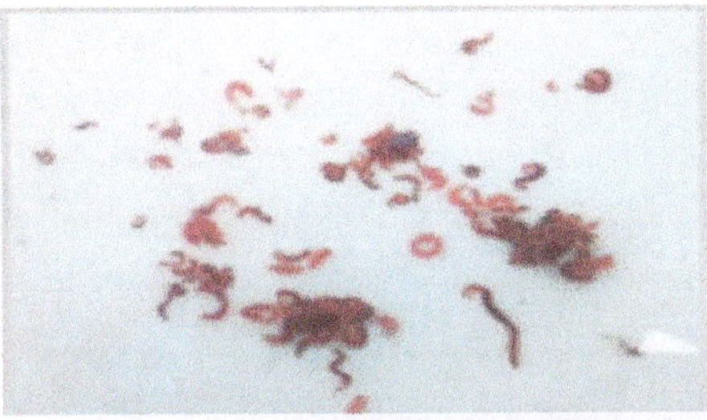

Figure 8.3. Culture, Reproduction and Life Cycle of Polychaete Worms, *Marphysa gravelyi*. (pg. 116)

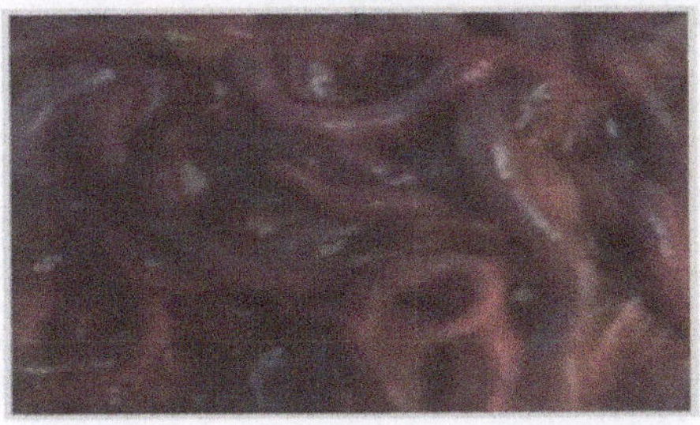

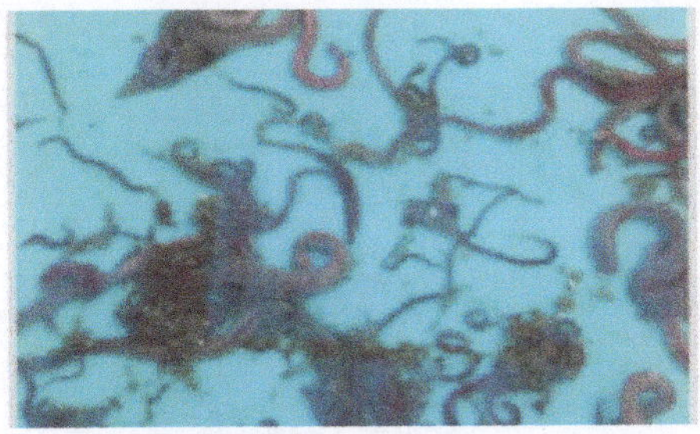

Figure 8.3. Culture, Reproduction and Life Cycle of Polychaete Worms, *Marphysa gravelyi*. (pg. 116)

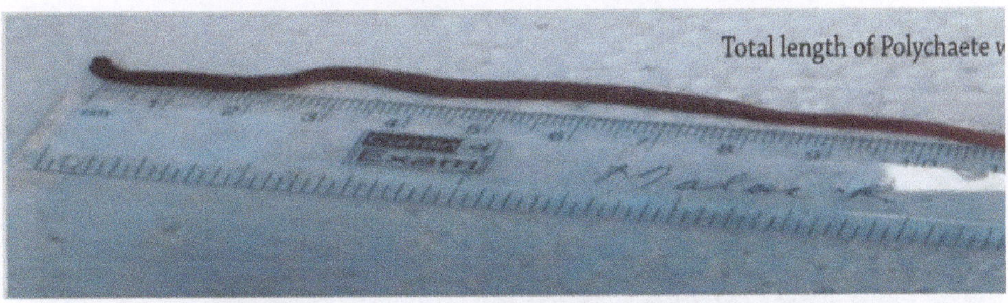

Figure 8.3. Culture, Reproduction and Life Cycle of Polychaete Worms, *Marphysa gravelyi*. (pg. 116)

Figure 8.3. Culture, Reproduction and Life Cycle of Polychaete Worms, *Marphysa gravelyi*. (pg. 117)

Figure 8.3. Culture, Reproduction and Life Cycle of Polychaete Worms, *Marphysa gravelyi*. (pg. 117)

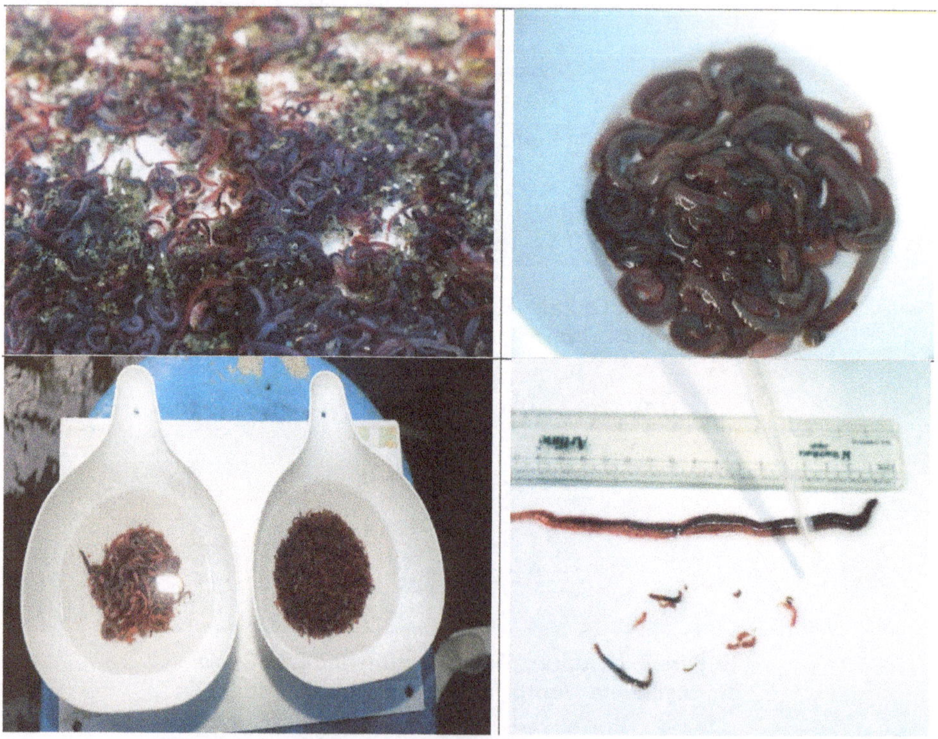

Figure 8.5. Grow-out Practices of Mud Polychaete Worm, *Marphysa gravelyi* in FRP Tanks. (pg. 118)

Figure 8.5. Grow-out Practices of Mud Polychaete Worm, *Marphysa gravelyi* in FRP Tanks. (pg. 118)

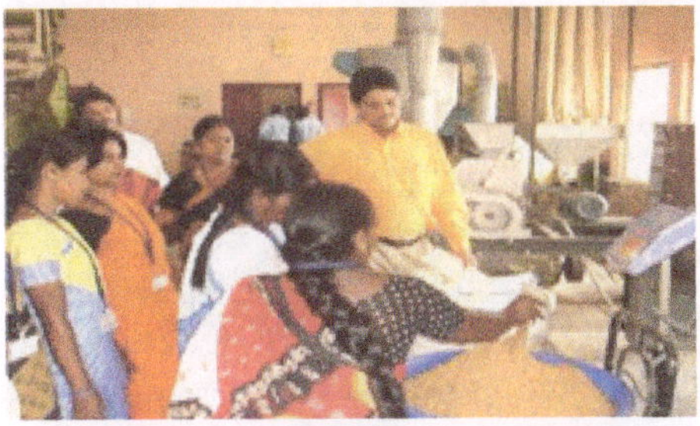

Figure 9.2. Weighing. (pg. 125)

Figure 9.2. Weighing. (pg. 126)

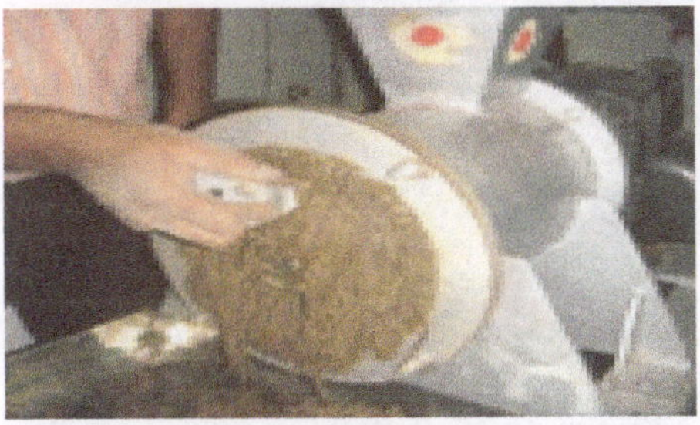

Figure 9.2. Weighing. (pg. 126)

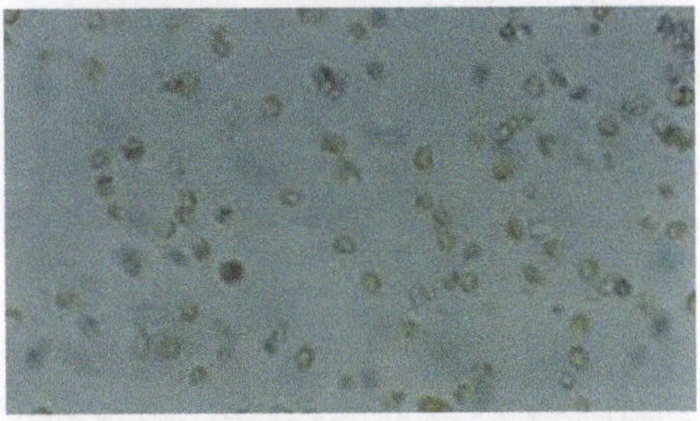

Figure 10.1. *Isochrysis galbana* (pg. 130)

Figure 10.2. *Nannochloropsis oculata* (pg. 130)

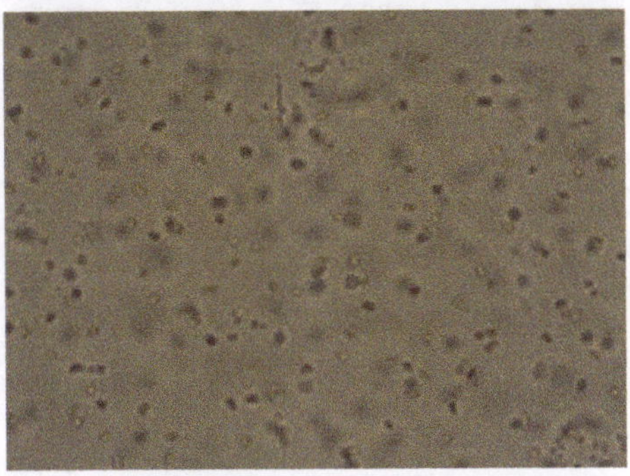

Figure 10.3. *Chaetoceros calcitrans* (pg. 131)

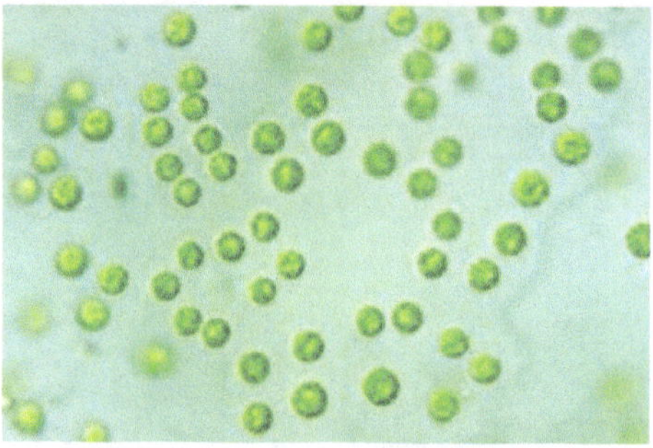

Figure 10.4. ***Chlorella salina*** **(*Source*: Google)(pg. 131)**

Figure 10.5. *Pavlova lutheri* **(pg. 131)**

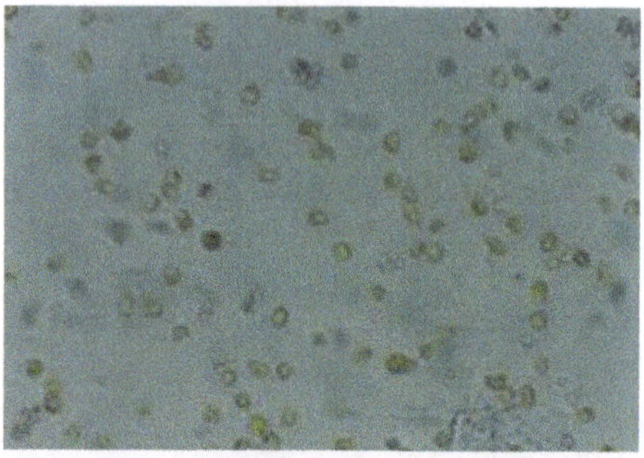

Figure 10.6. *Tetraselmis suecica* **(pg. 131)**

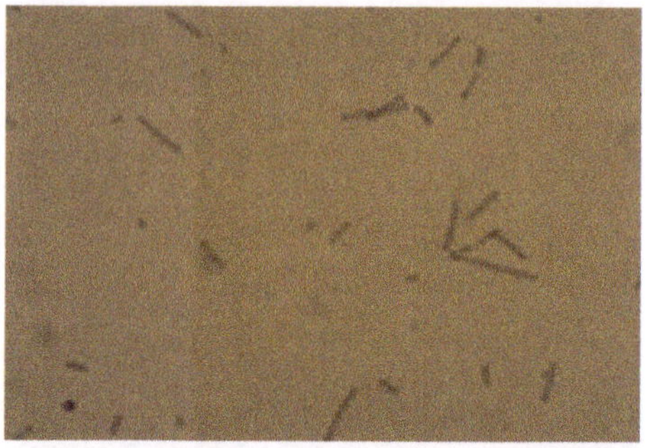

Figure 10.7. *Thalassiosira weissflogii* **(pg. 133)**

Figure 10.8. *Skeletonema costatum* **(pg. 133)**

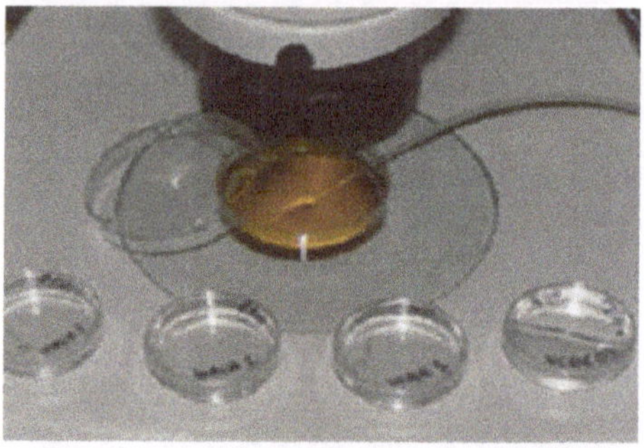

Figure 10.12. Isolation of Single Cell with the Help of Microscope. (pg. 138)

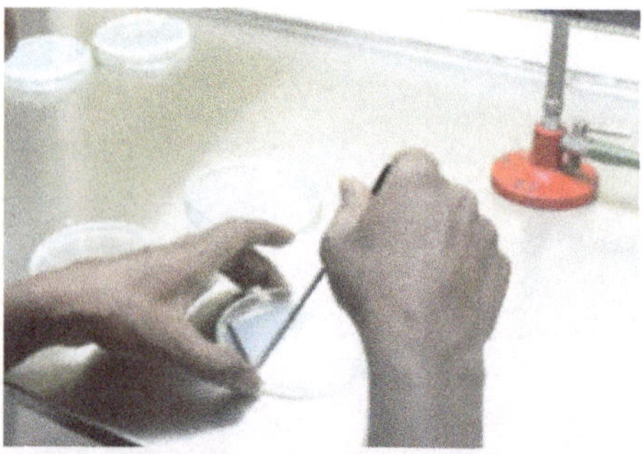

Figure 10.13a. Steaking of Micro Algal Culture on Agar Plates. (pg. 139)

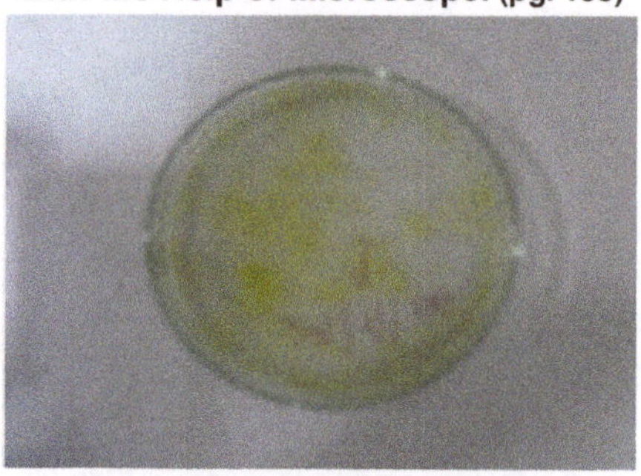

Figure 10.13b. Micro Algae (*Nannochloropsis* sp.) Colonies on in Agar Plates. (pg. 139)

Figure 10.14. Stock and Intermediate Culture of *Nannochloropsis oculata* and *Isochrysis galbana*. (pg. 140)

Figure 10.16. Outdoor Mass Culture of Micro Algae. (pg. 141)

Figure 10.17. Outdoor Mass Culture of Micro Algae in 1t FRP Tanks. (pg. 141)

Figure 10.18. Concentrated *Nannochloropsis oculata* Cells before and after Dilution (20X). (pg. 141)

Figure 10.20. Rotifer Culture on *Nannochloropsis oculata* Concentrate. (pg. 141)

Figure 10.21. Green Water Rearing System from Nannochloropsis. (pg. 141)

Figure 10.19. Diluted *Nannochloropsis oculata* Cells from Concentrate before and after Evan's Blue Staining. (pg. 145)

Figure 11.1: Indian Pompano. (pg. 147)

Figure 11.2. Broodstock Tank. (pg. 149)

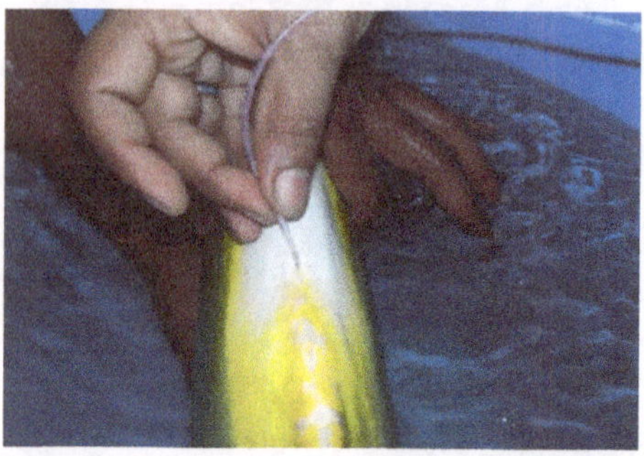

Figure 11.3. Cannulation of Indian Pompano using Baby Feeding Tube. (pg. 150)

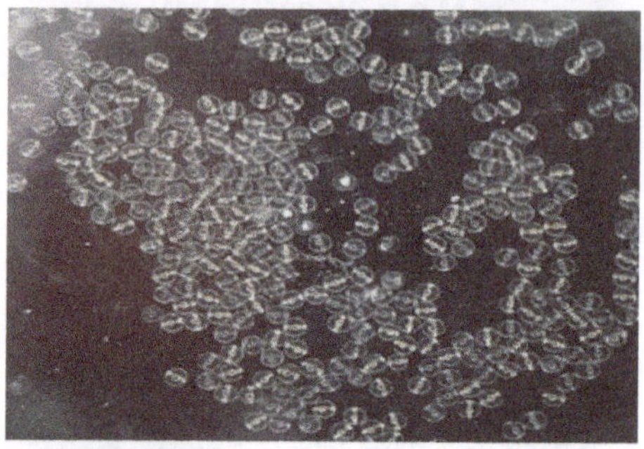

Figure 11.4. Fertilized Eggs of Indian Pompano, Embryo Development gives them a Clear View. (pg. 152)

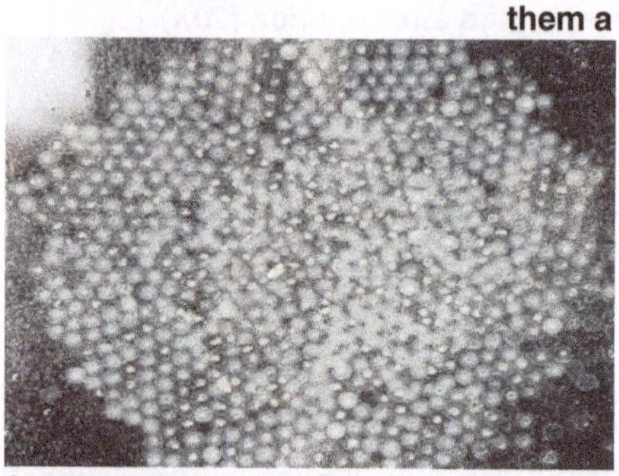

Figure 11.5. Unfertilized Eggs of Indian Pompano make them Opaque. (pg. 152)

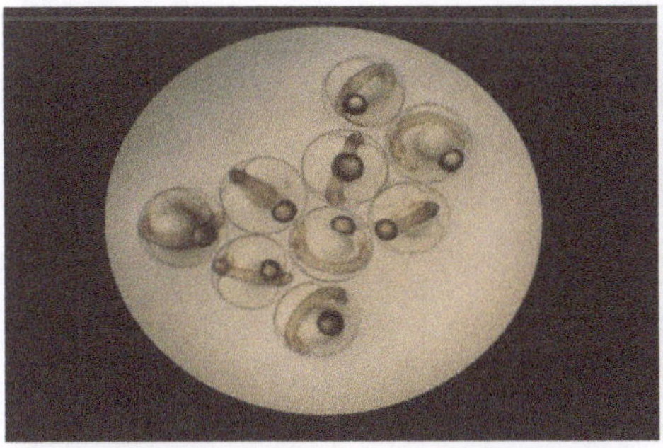

Figure 11.6. Eyed Stage Indian pompano Eggs Immediately before Hatching. (pg. 152)

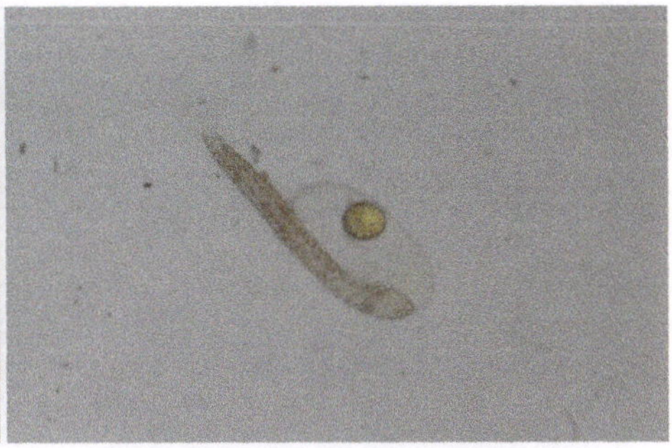

Figure 11.7. Newly Hatched Indian Pompano Larva. (pg. 152)

1st DPH

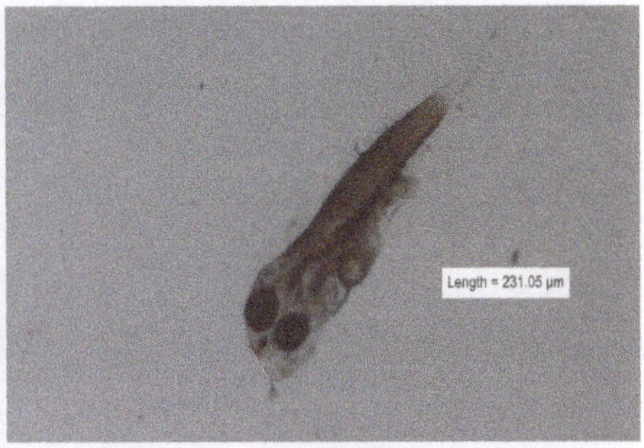

2nd DPH

3rd DPH

4th DPH

5th DPH

6th DPH

Figure 11.8: Larval Development Stages of Indian Pompano. (pg. 155-156)

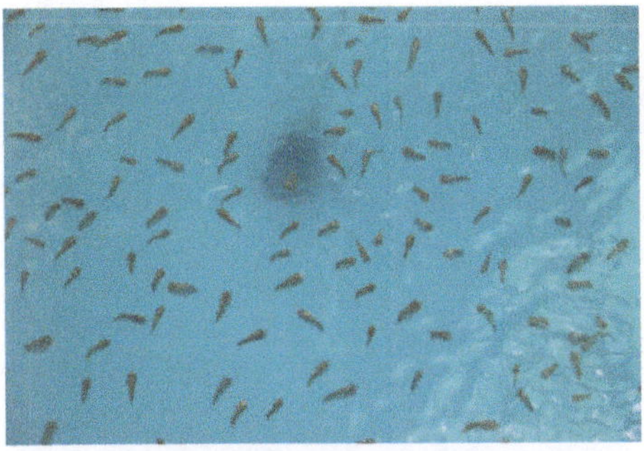

Figure 11.9. Metamorphosed Larvae of Indian Pompano. (pg. 152)

Figure 11.10. Orange Spotted Grouper *Epinephelus coioides*. (pg. 152)

Figure 11.11. Brooders *E. coioides*. (pg. 158)

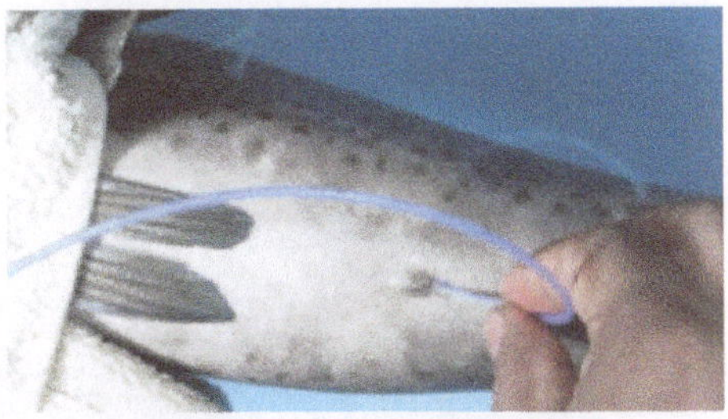

Figure 11.12. Cannulation of Organge Spotted Grouper using Baby Feeding Tube. (pg. 160)

Figure 11.14. Late-Stage (Eyed Stage) Orange Spotted Grouper Eggs Immediately before Hatching. (pg. 161)

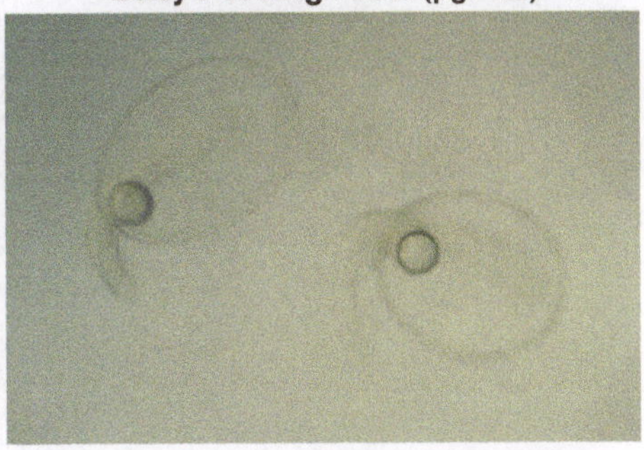

Figure 11.15. Newly Hatched Orange Spotted Grouper Larva. (pg. 161)

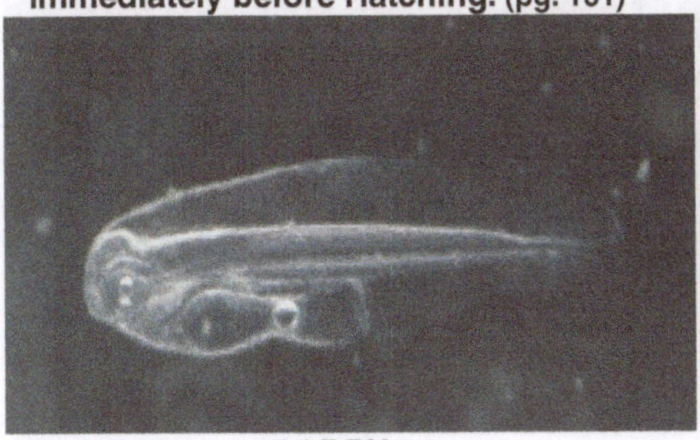

1st DPH

2nd DPH

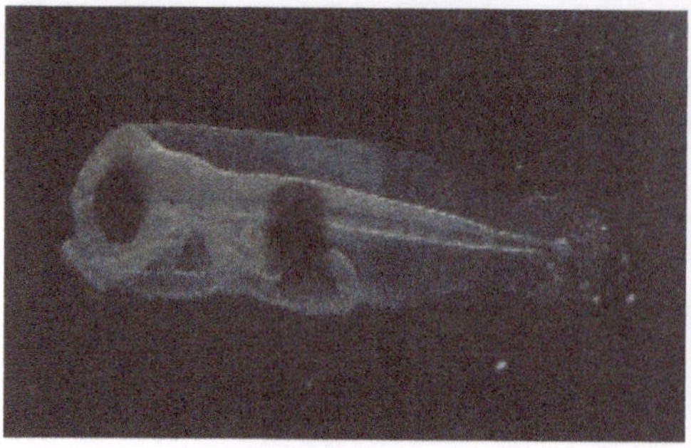

3rd DPH

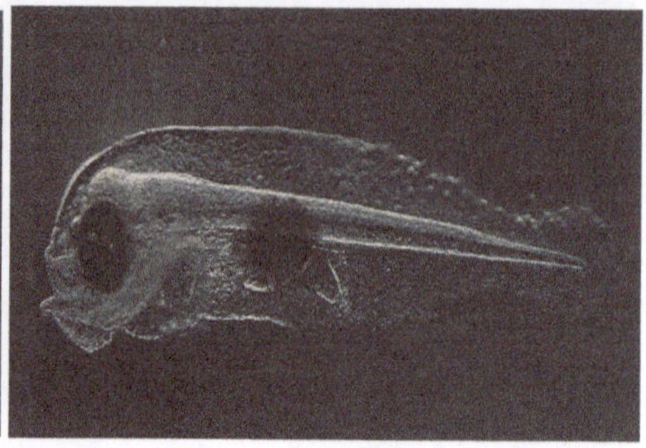

4th DPH

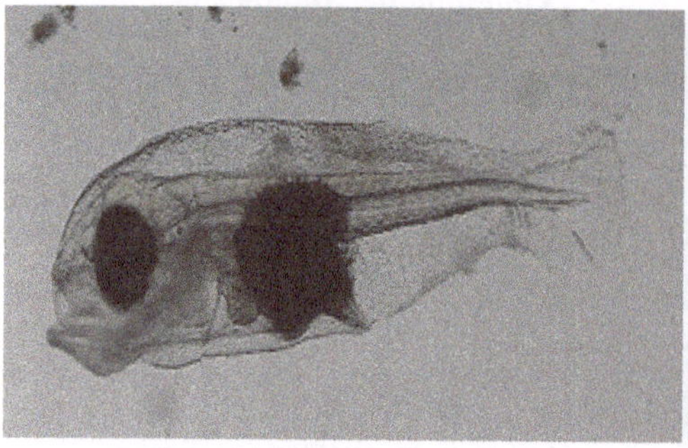

6th DPH

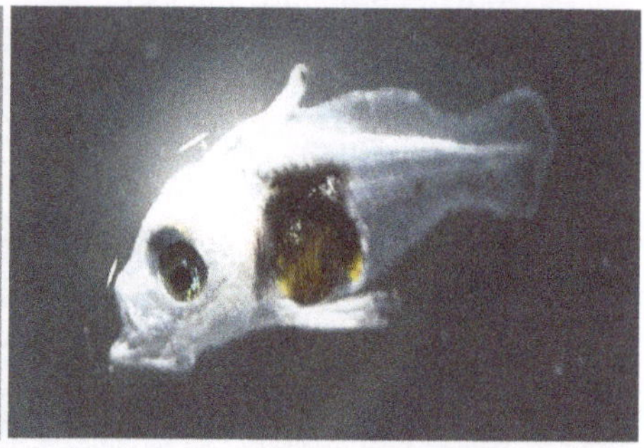

8th DPH

11th DPH

13th DPH

15th DPH

19th DPH

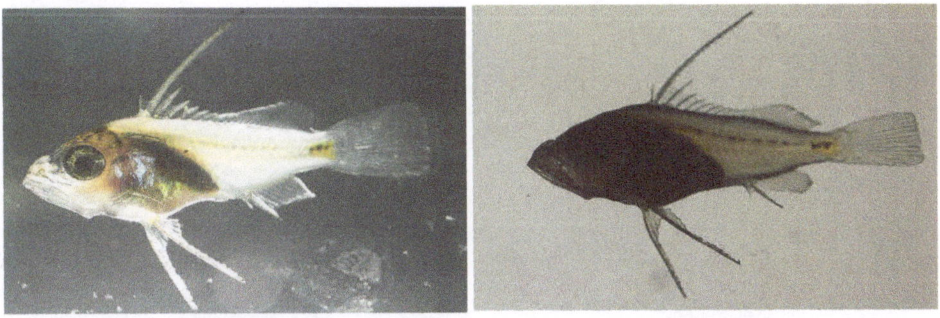

25th DPH 29th DPH

Figure 11.16. Larval Development Stages of Orange Spotted Grouper. (pg. 163-164)

Figure 11.17. Metamorphosed Larvae of Orange Spotted Grouper (*Epinephelus coioides*). (pg. 164)

Figure 12.1. Open Sea Cages. (pg. 173)

Figure 12.2. Indian Pompano in Sea Cage. (pg. 176)

Figure 12.3. Cage Culture of Orange Spotted Grouper. (pg. 177)

Figure 12.3a. Harvesting of Cage Cultured Orange Spotted Grouper. (pg. 178)

Figure 12.4. Grow-out Culture of Indian Pompana in Coastal Pond. (pg. 180

Figure 12.5. Nursery Culture of Orange Spotted Grouper in Coastal Ponds. pg. (181)

Figure 12.6. Grow-out Culture of Orange Spotted Grouper in Coastal Ponds. pg. (181)

Figure 13.6. Typical Layout of a Biofloc Farm. pg. (189)

Figure 13.3. Lined Pond Constructed Using 330 GSM SFF-HDPE Lining Materials. pg. (187)

Index

I

L

M

N

O

P

9 789359 191973

Printed at Repro India Ltd.